Selected Titles in This Series

Generalized Analytic Continuation

University
LECTURE
Series

Volume 25

Generalized Analytic Continuation

William T. Ross
Harold S. Shapiro

American Mathematical Society
Providence, Rhode Island

EDITORIAL COMMITTEE

2000 *Mathematics Subject Classification.* Primary 30B40, 47B38, 30B30, 30E10.

ABSTRACT. The theory of generalized analytic continuation studies continuations of meromorphic functions in situations where traditional theory says there is a *natural boundary*. This broader theory touches on a remarkable array of topics in classical analysis, as described in the book. The authors use the strong analogy with the summability of divergent series to motivate the subject. They are careful to cover the various types of continuations, attempting to unify them and suggesting some open questions. The book also addresses the role of such continuations in approximation theory and operator theory. The introductory overview provides a useful look at the history and context of the theory.

Library of Congress Cataloging-in-Publication Data

Ross, William T., 1964–
    Generalized analytic continuation / William T. Ross, Harold S. Shapiro.
       p. cm. — (University lectures series, ISSN 1047-3998 ; v. 25)
    Includes bibliographical references and index.
    ISBN 0-8218-3175-5 (alk. paper)
    1. Analytic continuation. I. Title. II. Series: University lecture series (Providence, R.I.) ;
25.

QA331.7.R68   2002   2002018463
515′.9–dc21   CIP

Contents

Preface

In these notes, we shall present a body of mathematical work devoted to generalizing the classical Weierstrassian notion of analytic continuation. These works by Poincaré, Borel, Beurling, Tumarkin, Gončar, and others have been undertaken by various methods and for a variety of reasons, but for brevity, we will refer to them as studies in 'generalized analytic continuation' (GAC). To better explain the motivation behind our assembling this little book, we employ an analogy with the, now classical, subject of divergent series.

In the 18th century, there was little agreement on how to attach an appropriate number s to represent the 'sum' $a_0 + a_1 + a_2 + \cdots$ of complex numbers a_i. From G. H. Hardy's *Divergent Series* [**68**, p. 5]:

> It is plain that the first step towards such an interpretation must be some definition, or definitions, of the 'sum' of an infinite series,. . . . This remark is trivial now: it does not occur to a modern mathematician that a collection of mathematical symbols should have a 'meaning' until one has been assigned to it by definition. It was not a triviality even to the greatest mathematicians of the eighteenth century. They had not the habit of definition: it was not natural to them to say, in so many words, 'by X we *mean* Y'. There are reservations to be made... but it is broadly true to say that mathematicians before Cauchy asked not 'How shall we *define* $1 - 1 + 1 - 1 + \cdots$?' but 'What *is* $1 - 1 + 1 - 1 + \cdots$?', and that this habit of mind led them into unnecessary perplexities and controversies which were often really verbal.

Some order was established in this regard when Cauchy first defined, in a way that is now standard, the notion of a 'limit' of a sequence, allowing him to define the sum $a_0 + a_1 + a_2 + \cdots$ to be the number s if

$$\lim_{n \to \infty} (a_0 + a_1 + \cdots + a_n) = s.$$

This definition banished series such as $1 - 1 + 1 - 1 + \cdots$ to the realm of 'divergent' series.

However, various investigations pointed towards the usefulness of certain divergent series. From James Pierpont:

> It is indeed a strange vicissitude of our science that those series which early in the century were supposed to be banished once and for all from rigorous mathematics should, at its close, be knocking at the door for readmission.

For example, the initial treatment (by Fourier) of Fourier series operated extensively with divergent series. As another example, important series expansions arising in celestial mechanics - and also in connection with the Euler-Maclaurin summation formula - turned out to diverge, yet gave useful results if truncated suitably, a phenomenon later explained by Poincaré and others in the theory of asymptotic series. Still further, Euler, Borel, and Mittag-Leffler had devised procedures which 'summed' certain Taylor expansions in regions outside their circle of convergence to the value of the analytically continued function [**75**, p. 74]. Even further, after the disappointing discovery, by du Bois-Reymond, that the Fourier series of a continuous function could diverge at some points, order was restored in a remarkable way by L. Fejér's discovery that none the less, the arithmetic means of those partial sums do converge uniformly to the function. Lebesgue proceeded further and showed that the same arithmetic means formed from the partial sums of the Fourier series of a merely integrable function converge almost everywhere to the function. These are only some examples. A more detailed discussion can be found in [**68**].

As various methods of 'summing' divergent series proved their worth, and also became systematized by Toeplitz and others, a new 'modern' point of view emerged: there was no need for polemics as to what was the 'correct' sum of a series like $1 - 1 + 1 - 1 + \cdots$ Indeed, various reasonable looking summation procedures lead to different answers. Instead, what was truly needed was a classification and a comparison of the various summation procedures of Abel, Cesàro, Borel, and Euler, and an even greater emphasis on their applications.

Among the new challenges in the field of divergent series was the task of proving that the various new types of summation were compatible with Cauchy's definition of the sum in the case where the series was convergent in the first place. These results are generally referred to as 'Abelian' theorems since the first was a result of Abel asserting that if $a_0 + a_1 + a_2 + \cdots$ converges to s then

$$\lim_{x \to 1^-} \sum_{n=0}^{\infty} a_n x^n = s,$$

with the above formula being the 'Abel sum' of the a_i's. A similar type of result holds for Cesàro convergence, that is

$$\lim_{n \to \infty} \frac{s_0 + s_1 + \cdots + s_n}{n+1} = s,$$

where $s_n = a_0 + a_1 + \cdots + a_n$, and the above formula being the 'Cesáro sum' of the a_i's. Another important task was to explore the converse question, and more generally the compatibility of different types of summability - what are referred to as 'Tauberian' theorems [**68**, p. 149]. For example, a result of Tauber says that if a sequence $\{a_n\}$ is Abel summable to s, as above, and $a_n = o(1/n)$, then $a_0 + a_1 + \cdots = s$.

Though there was initially little unity in the subject of divergent series, since the motivations as well as the methods were diverse, unifying studies began to appear in the work of Toeplitz and others. Finally, a comprehensive synthesis of the whole field, including historical background, was done by G. H. Hardy in his masterful work *Divergent Series* [**68**].

In much the same way as Euler, Abel, Borel, Cesàro and others attempted to extend the notion of convergence of a series, there were those who attempted to extend the notion of analytic continuation of a function f. In brief, 'generalized analytic continuation' (GAC), as the name suggests, studies ways in which the component functions $f|\Omega_j$ - where f is a meromorphic function on a disconnected open set Ω in the complex plane $\mathbb{C}$ and $\{\Omega_j\}$ are the connected components of Ω - may possibly be related to each other in certain cases where the Weierstrassian notion of analytic continuation says there is a 'natural boundary'. We say that $\partial\Omega_j$, the boundary of Ω_j, is a 'natural boundary' for the component function $f|\Omega_j$ if $f|\Omega_j$ does not have an analytic continuation across any point of $\partial\Omega_j$. As an example of what is meant here, consider the function f defined by the series

$$f(z) = \sum_{n=1}^{\infty} \frac{c_n}{z - e^{i\theta_n}},$$

where $\{e^{i\theta_n}\}$ is a sequence of points everywhere dense in the unit circle $\mathbb{T} = \{z \in \mathbb{C} : |z| = 1\}$ and $\{c_n\}$ is an absolutely summable sequence of non-zero complex numbers. This function is analytic on $\mathbb{C}_\infty \backslash \mathbb{T}$ and in 1883, Poincaré [106] was able to prove that $\mathbb{T}$ is a natural boundary for both $f|\mathbb{D}$ and $f|\mathbb{D}_e$, where $\mathbb{D} = \{z \in \mathbb{C} : |z| < 1\}$ is the unit disk and $\mathbb{D}_e = \{z \in \mathbb{C}_\infty : 1 < |z| \leq \infty\}$ is the extended exterior disk.

Certainly, if Ω_j and Ω_k are adjacent components (they share a common boundary arc) and $f|\Omega_j$ is an analytic continuation of $f|\Omega_k$ across a portion of that common boundary arc, then $f|\Omega_j$ and $f|\Omega_k$ are related, and, by the uniqueness of analytic continuation, each uniquely determines the other. However, there are other types of 'continuations', or ways of relating $f|\Omega_j$ and $f|\Omega_k$, beyond analytic continuation. From É. Borel's work [28, p. 100], where such ideas began to be studied:

> ... we wished only to show how one could introduce into the calculations analytic expressions whose values, in different regions of their domain of convergence, are mutually linked in a simple way. It seems, on the basis of that, that one could envision extending Weierstrass' definition of *analytic function* and regarding in certain cases as being [parts of] *the same function*, analytic functions having separate domains of existence. But for that it is necessary to impose restrictions on the analytic expressions one considers, and because he did not wish to impose such restrictions Weierstrass answered in the negative [this] question:
>
> "Therefore the thought was not to be ignored, as to whether in the case where an arithmetic expression $F(x)$ represents different monogenic functions in different portions of its domain of validity, there is an essential connection, with the consequence that the properties of the one should determine the properties of the other. Were this the case, it would follow that the concept 'monogenic function' must be widened."- (Weierstrass, *Mathematische Werke*, vol. 2, p. 212)
>
> It is not possible for us to give to this Chapter a decisive conclusion; for, in our opinion, the question addressed here is not entirely resolved and calls for further research. We would be content if we have convinced our readers that neither the fundamental works of Weierstrass, nor the later ones of Mittag-Leffler, Appell, Poincaré, Runge, Painlevé entirely answer the question

as to the relations between the notions of analytic function, and analytic expression. One can even say without exaggeration, that the classification of analytic expressions which are incapable of representing zero [on some domain] without doing so everywhere, is yet to be brought to completion.

For an example of a 'continuation' beyond analytic continuation, we return to the above Poincaré example

$$f(z) = \sum_{n=1}^{\infty} \frac{c_n}{z - e^{i\theta_n}}.$$

Even though the component functions $f|\mathbb{D}$ and $f|\mathbb{D}_e$ are not analytic continuations of each other across any point of $\mathbb{T}$, they can, in light of more modern complex analysis techniques, be regarded as 'continuations' of each other. Indeed, $f|\mathbb{D}$ and $f|\mathbb{D}_e$ are both in the H^p $(0 < p < 1)$ spaces of their respective domains and moreover,

$$\lim_{r \to 1^-} f(re^{i\theta}) = \lim_{r \to 1^+} f(re^{i\theta})$$

for almost every θ. Borel [**27, 28**] began to generalize Poincaré's example to functions of the type

$$f(z) = \sum_{n=1}^{\infty} \frac{c_n}{z - z_n},$$

where $\{z_n\}$ is a sequence of points which need not *lie* on $\mathbb{T}$ but merely *accumulate* on all of $\mathbb{T}$, and further explored the relationship between the component functions $f|\mathbb{D}$ and $f|\mathbb{D}_e$. This investigation of 'coherence', often in various other settings, was continued in the 1920's and 1930's by Walsh and more recently in the 1960's and 1970's by one of the current authors as well as Tumarkin and Gončar.

An important insight emerged from these pioneering works: if a sequence of rational functions converges uniformly on compact subsets of a disconnected open set - as the partial sums of the above series do - and the approximants are restricted by either of two very different types of requirements, then the limit functions on the components of the open set exhibit a kind of mutual 'coherence'. In particular, they mutually determine one another. These two types of requirements are, on the one hand, a sufficiently rapid convergence as related to the degree of the rational function (a line pursued by Borel, Walsh, and Gončar), and on the other, a geometric restriction of the locations of the poles, without regard to the speed of convergence (a line pursued by Walsh and Tumarkin). One of the current authors and his co-workers later revealed a surprising connection between Tumarkin's results and a problem in operator theory, namely the classification of the cyclic vectors for the backward shift operator on the Hardy space.

The results we have assembled in this little book possess a mathematical richness and potential to constitute an interesting field in much the same way as divergent series. But we are not there yet. To again pursue the analogy with divergent series, so far we have a variety of proposed schemes for continuing (classes of) noncontinuable analytic functions and 'Abelian' theorems, guaranteeing they will produce the correct result when applied to functions that already possess ordinary analytic continuations. But the analog of the Tauberian theorems is almost wholly lacking. For example, it is not known, at least to us, whether pseudocontinuation and Gončar continuation, as defined in this book, applied to the same function can yield incompatible results. There is also another interesting analogy with divergent series. Just as the Tauberian theorems showed that some divergent series, for example, $1/2 + 1/3 + 1/4 + \cdots$, are, roughly speaking, 'universally divergent' (they cannot be summed by any of the natural methods), some Taylor series, such as the ones which have an isolated winding point on their circle of convergence, are 'universally non-continuable' and do not allow a GAC by any method satisfying a few natural conditions. This limitation of the scope of GAC is a sign that we are dealing with a *bona fide* concept here since it cannot be stretched to make 'everything' continuable.

It is our hope that the present book, which juxtaposes results in a way that has not been done before, may help pave the way towards the study of GAC in a unifying context. We feel our pursuit is worthwhile since for one, much of the mathematics involved here is quite beautiful but unfortunately widely scattered throughout the literature, often in sources difficult to obtain. Secondly, it involves a central notion in function theory, analytic continuation, which historically has given rise to heated polemics. In the same spirit as the modern theory of summability methods for divergent series, there is nothing controversial here, only technical questions of how different continuation schemes relate to one another and to analytic continuation. Thirdly, many attributes of analytic continuation can be studied afresh in the context of one or another notion of GAC, such as the overconvergence theorems of Ostrowski or noncontinuability of Taylor series with various types of gaps. Fourth, GAC appears in one form or another in the study of the backward shift operator on many Banach spaces of analytic functions on the unit disk. We have already mentioned this with regard to the Hardy space. Recently however, several authors have employed GAC in their investigations of the backward shift on other function spaces such as the Bergman and Dirichlet spaces. In fact, a new type of 'continuation' to be introduced in these notes leads to further progress in this context. Finally, GAC appears, almost surprisingly, in the study of electrical networks (the Darlington synthesis problem) as well as in questions related to linear differential equations of infinite order.

Though the subject of GAC is not ready for a *Divergent Series* like treatise, we feel this humble offering begins to organize the results in a coherent way and presents the future author of such a treatise with some open questions that need to be answered before such a comprehensive work can be written.

We wish to give our warmest thanks to Dima Khavinson, Genrikh Tumarkin, and Lawrence Zalcman who read earlier drafts of this manuscript and provided valuable suggestions and corrections. We also thank Evgeny Abakumov for introducing us to Aleksandrov's papers on gap series.

We will be keeping any updates/corrections/additions at

`www.richmond.edu/~wross`

We welcome your comments.

W. T. Ross   H. S. Shapiro
Richmond   Stockholm
wross@richmond.edu   shapiro@math.kth.se

CHAPTER 1

Overview

There has often been confusion during the development of our modern concept of a 'function'. Originally, one clung to the naive idea of a function as being given by a 'formula', what we today call a rational function. Later, based on new concepts from infinitesimal analysis, infinite sums and integrals were permitted into the formulae to describe functions. This led to apparent paradoxes when it was found, for example, that a reasonably nice Fourier cosine series could sum to the value zero on one interval and the value one on another. In other words, one formula represented two seemingly different functions. In the realm of analytic functions, one observed similar 'one formula - two functions' phenomena, for example,[1]

$$(1.0.1) \qquad \frac{z}{1-z^2} + \frac{z^2}{1-z^4} + \cdots + \frac{z^{2^{n-1}}}{1-z^{2^n}} + \cdots = \begin{cases} \dfrac{z}{1-z} & \text{if } |z| < 1, \\[2ex] \dfrac{-z}{1-z} & \text{if } |z| > 1. \end{cases}$$

As our understanding of functions evolved, the old idea of 'one formula - one function' seemed to fade into oblivion as we became more comfortable with the concept that one formula could represent several seemingly unrelated functions on different components of its domain of definition. However, some analysts kept the idea of 'one formula - one function' alive into the modern era, not so much out of stubbornness, but, thanks to the advances made in the foundations of complex analysis, the issue could be raised to a higher level of sophistication than the almost trite example in eq.(1.0.1).

The main way of associating or relating the functions $f|\Omega_j$, where f is an analytic (or meromorphic) function on a disconnected open set $\Omega \subset \mathbb{C}$ and $\{\Omega_j : j = 1, 2, \ldots\}$ are the connected components of Ω, uses the notion of analytic continuation, according to Weierstrass, as defined in all modern complex analysis textbooks. However, there are examples of analytic functions in the unit disk $\mathbb{D}$ defined by a power series, for example

$$\sum_{n=0}^{\infty} z^{2^n},$$

which can not be analytically continued across any point of the unit circle $\mathbb{T} = \partial\mathbb{D}$. That is to say, $\mathbb{T}$ is a 'natural boundary' or *coupure* (cut) for the function and it is totally meaningless to speak of this function outside the disk.

While this idea was logically unassailable, some analysts felt it was too restrictive, since, unlike power series, which either have a disk or the whole plane as the

---

[1]To verify this identity for $|z| < 1$, write each of the fractions as a Taylor series. For $|z| > 1$, write $w = 1/z$ and observe the identity $\dfrac{z^m}{1-z^{2m}} = \dfrac{w^m}{w^{2m}-1}$.

maximal open set of convergence, other types of series or limiting procedures could very well converge in sets which disconnect the plane. Indeed eq.(1.0.1) is an example of this. With this in mind, one can ask whether or not the component functions of a sum of a series, or perhaps the result of some other limiting procedure, should be considered as one, rather than several functions. This led, quite naturally, to the concept of 'generalized analytic continuation', which, as its name suggests, tries to imagine other ways in which $f|\Omega_j$ and $f|\Omega_k$, for adjacent components Ω_j and Ω_k of the domain of definition of f, can be thought of as 'continuations' of each other even though they are not *bona fide* analytic continuations.

To be sure, the function defined in eq.(1.0.1) fails in this regard, since its limiting values in the interior and in the exterior of the unit circle, the two components of its domain of definition, can by no stretch of the imagination be considered as two parts of a coherent whole. On the other hand, in 1883, Poincaré [**106**] drew attention to the function f defined by the series

$$(1.0.2) \qquad\qquad f(z) = \sum_{n=1}^{\infty} \frac{c_n}{z - z_n},$$

where $\{c_n\}$ is an absolutely summable sequence of non-zero complex numbers and $\{z_n\}$ is a sequence of points on a smooth closed curve L which are everywhere dense in L. It is routine to show that the series converges on the complement, L^c, of L and that f defines an analytic function on each of the two components of L^c, namely $\mathrm{int}(L)$ (the interior of L) and $\mathrm{ext}(L)$ (the exterior of L). Poincaré showed, in the case where L bounds a convex region, that L is a *coupure* for the functions $f|\mathrm{int}(L)$ and $f|\mathrm{ext}(L)$ (see Chapter 3). These results were later extended by Denjoy [**43**] in 1926 to allow L to be an arbitrary Jordan curve. The importance of Poincaré's result, as interpreted by his teacher Hermite, was that it gave a new construction of a nowhere continuable analytic function, at that time, a subject still arousing intense interest. In slightly different terms, the 'apparent singularities' $\{z_n\}$ really were singular points, as opposed to those in eq.(1.0.1) which evaporate in the limit.

So, for the function in eq.(1.0.2), are we talking about one function or two? In light of a concept not yet known in Poincaré's time, one can go further and say that these two component functions have equal non-tangential limiting values at almost all points of L (see Chapter 3). In view of a 1925 theorem of Lusin and Privalov (see Theorem 2.2.2), this implies that $f|\mathrm{int}(L)$ and $f|\mathrm{ext}(L)$ uniquely determine one another, and, despite the fact that they are not analytic continuations of each other, one can say that $f|\mathrm{ext}(L)$ is a 'continuation' of $f|\mathrm{int}(L)$. In 1966, this Poincaré example became the model for the definition of 'pseudocontinuation' [2] by one of the present authors [**128**], and will be much discussed throughout these notes.

From Borel's point of view [**27, 28**], the series in eq.(1.0.2) can be seen as defining a single function on a domain consisting of the whole complex plane minus a certain exceptional set on L, which we nowadays would call a set of linear measure zero, and, which is continuous along every line wholly in this domain and not tangent to L. We shall not pursue this point of view here but mention that A. Denjoy considered Borel's ideas about *fonctions monogènes* historically important for at least two reasons: (i) it foreshadowed what was later to become the theory of quasi-analytic functions of a real variable, which also came into focus as a result

---

[2]*i.e.*, the component functions have equal non-tangential limits almost everywhere (see Definition 6.2.1 for a more precise definition and for further examples).

of Hadamard's work on partial differential equations [**66**]; and (ii) the notion 'set of measure zero' was introduced into analysis in this connection, in advance of Lebesgue's theory of measure.

Borel tried to extend the class of examples introduced by Poincaré by choosing $\{z_n\}$ not on a curve, but only *clustering* (everywhere) on such a curve, or more generally on a set L separating the plane. Again, there was the question of the coherence properties of the component functions of

$$\sum_{n=1}^{\infty} \frac{c_n}{z - z_n}.$$

For example, if one were going to associate in some way the component functions of the above series, one would hope to show the following 'Abelian' theorem: denoting by $\{\Omega_j : j = 1, 2, \ldots\}$ the connected components of L^c, if the restriction $f|\Omega_j$ is analytically continuable along some path into Ω_k, that continuation must be given on Ω_k by the sum of the series, *i.e.*,

$$\sum_{n=1}^{\infty} \frac{c_n}{z - z_n}.$$

We use the phrase 'Abelian' theorem to refer to the property that a new type of 'continuation', as alluded to above, should not be incompatible with the time honored notion of analytic continuation. The analog here is to the Abelian theorems of divergent series (as mentioned in the 'Preface') where one would hope that any new type of definition of a 'sum' of an infinite series should not come into conflict with the generally accepted Cauchy notion of convergence. This Abelian theorem Borel was able to show [**27**], but only under the hypothesis

$$(1.0.3) \qquad\qquad \lim_{n \to \infty} \sqrt[n]{|c_n|} = 0.$$

Eventually, Gončar showed (see below) that this condition could be relaxed, in the case of two adjacent domains, even further to

$$\overline{\lim_{n \to \infty}} \sqrt[n]{|c_n|} < 1.$$

It was a long open question as to whether or not the above Abelian theorem holds for such a series [3], under only the weaker and very natural hypothesis that the sequence $\{c_n\}$ is absolutely summable (as in the Poincaré example). In 1921, J. Wolff [**146**] found a counterexample which, after amelioration by others such as Bonsall [**26**], Denjoy [**42, 43**], Beurling [**18, 22**], and Leont'eva [**94**], showed that a Borel series can sum to zero on just a single component of L^c, even with very rapid, but not exponential, decay of the coefficients $\{c_n\}$. More recently, a paper of Brown, Shields, and Zeller [**32**] gave a necessary and sufficient geometric condition on a sequence of poles $\{z_n\} \subset \mathbb{D}$ so that one can find an absolutely summable sequence of non-zero complex numbers $\{c_n\}$ whose associated Borel series

$$\sum_{n=1}^{\infty} \frac{c_n}{z - z_n}$$

sums to zero on the exterior disk. A thorough discussion of Borel series and their continuation properties will be taken up in Chapter 4, § 4.2.

---

[3]We shall henceforth call them 'Borel series', even though Borel was not the first to study them.

J. L Walsh [143] generalized Borel series, with rapidly decreasing coefficients, to meromorphic functions f which are the limits of very rapidly convergent sequences of rational functions, what we shall call 'superconvergent' and 'hyperconvergent' sequences (see Definition 4.1.2 and Definition 5.1.3) and proved similar 'Abelian' results to those of Borel. For example, if $f|\Omega_j$ has an analytic continuation along some path to a point in Ω_k (for adjacent Ω_j and Ω_k), this continuation must agree with $f|\Omega_k$. Gončar (see Chapter 5) [62, 63] later extended this work of Walsh. It is interesting to note that although the basic techniques needed had been developed by Walsh, he did not seem to be looking for generalized methods of continuing analytic functions, though his theorems on 'overconvergence' of rapidly convergent sequences of rational functions certainly pointed in this direction. It was Gončar who built up a concept of 'generalized analytic continuation' on this basis.

As an illustration of generalized analytic continuation in Gončar's sense, we may consider the following modification of eq.(1.0.1):

$$p_1\frac{z}{1-z^2} + p_2\frac{z^2}{1-z^4} + \cdots + p_n\frac{z^{2^{n-1}}}{1-z^{2^n}} + \cdots = \begin{cases} f^+(z) & \text{if } |z| < 1, \\ f^-(z) & \text{if } |z| > 1. \end{cases}$$

where $\{p_n\}$ is a sequence of (non-zero) complex numbers decaying to zero exponentially. In this case, neither f^+ nor f^- are not analytically continuable across any point of the unit circle, yet are 'continuations' of one another by Gončar's definition. This coherence of f^+ and f^- is easy to understand even without the machinery of Gončar's method. The reader can check that in fact, by using a partial fraction decomposition on each term of the series and then gathering up terms with like poles, one can transform this function into a Poincaré-type example as in eq.(1.0.2). Hence f^+ and f^- are not only 'Gončar continuations' of each other but are also pseudocontinuations of each other, in the sense on Definition 6.2.1.

Another example falling under the Gončar definition of generalized analytic continuation is given by the partial sums of the Borel series

$$\sum_{n=1}^{\infty} \frac{c_n}{z - z_n}, \quad |z_n| \to 1,$$

but with the condition in eq.(1.0.3) relaxed to

$$\varlimsup_{n\to\infty} \sqrt[n]{|c_n|} < 1.$$

Here it is important to note that Walsh obtained (implicitly) and Gončar obtained (explicitly) the same 'Abelian' conclusion as Borel but with a less restrictive hypothesis on the decay of the $|c_n|$. In this case, exponential decay suffices.

Another side of the subject of generalized analytic continuation was opened by Walsh who, in 1929, investigated the following problem (focusing here on a special case and using modern terminology): For each $n \in \mathbb{N}$, let

$$S_n = \{ z_{n,1}, z_{n,2}, \ldots, z_{n,N(n)} \}$$

be a finite sequence of points from the extended exterior disk $\mathbb{D}_e$, where it is permitted that a point may appear more than once. Let

$$R_n := \bigvee \left\{ \frac{1}{z - z_{n,j}} : j = 1, \ldots, N(n) \right\}.$$

Here $\bigvee$ denotes the (closed) linear span. Note that if $z_{n,j} \neq \infty$ appears k times in S_n, then the spanning set for R_n is augmented to include the rational functions

$$\frac{1}{(z - z_{n,j})^s}, \quad s = 1, \ldots, k.$$

If $z_{n,j} = \infty$ appears k times in the sequence, then R_n includes the functions $1, z, \ldots, z^{k-1}$ in its spanning set. Walsh's question was: under what conditions is *every* function in the Hardy space H^2 of the unit disk $\mathbb{D}$ [4] representable as

(1.0.4) $$f = \lim_{n \to \infty} f_n,$$

where $f_n \in R_n$ for $n = 1, 2, \ldots$? Here the limit is in the H^2 norm. He showed this is so if and only if

$$p_n := \sum_{j=1}^{N(n)} \left(1 - \frac{1}{|z_{n,j}|} \right) \to \infty \ \text{ as } n \to \infty.$$

The question Walsh did not settle was: what can one say about the functions $f \in H^2$ satisfying eq.(1.0.4) in the case where

(1.0.5) $$\lim_{n \to \infty} p_n < \infty?$$

This question was completely settled in definitive papers of G. Tumarkin, especially [140], which cover a much more general field than we are taking up here. One of his results says that if one assumes eq.(1.0.5) holds, then every H^2 function f satisfying eq.(1.0.4) has the following remarkable 'continuation' property: there exists a meromorphic function g of 'bounded type' in $\mathbb{D}_e$ (*i.e.*, can be written as the quotient of two bounded analytic functions) whose non-tangential boundary values on the unit circle $\mathbb{T}$ are equal almost everywhere to those of f. Conversely, if f is an element of H^2 for which there exists a g of bounded type in $\mathbb{D}_e$ whose boundary values match those of f (almost everywhere), then there is a tableau $\{S_n\}$ as above, for which the sequence $\{p_n\}$ is bounded and such that f can be represented as in eq.(1.0.4). For example, the function $f = \log(1 - z)$ does not have this continuation property (Example 6.2.3), while a Blaschke product does (Example 6.2.4). This seems to us a very important discovery, probably the first time that the issue of a function possessing what we today call 'pseudocontinuation' (and, we stress, this may occur for f for which the unit circle is a *coupure*, in the sense of Weierstrass) turned up in an investigation not *a priori* concerned with matters of analytic continuation.

Thus, one can identify two distinct lines of investigation placing on the order of the day, so to speak, generalized analytic continuation, and each has to do with an aspect of rational approximation: In the line Borel, Walsh, and Gončar, the poles are quite freely placed, but strong restrictions are imposed on the residues (*i.e.*, series coefficients) to obtain 'coherent' functions. Whereas, in the line Walsh, Tumarkin, and others, one places no restrictions on the closeness of the approximations, but restricts the location of the poles in that they must stick 'close' to the unit circle.

---

[4]H^2 is the Hilbert space of analytic functions $f = \sum a_n z^n$ on $\mathbb{D}$ with norm defined by $\|f\|^2 = \sum_{n=0}^{\infty} |a_n|^2$.

The type of pseudocontinuation first encountered by Tumarkin has turned up frequently in recent investigations. For one thing, the investigation of cyclic vectors for the backward shift operator

$$Bf := \frac{f - f(0)}{z}$$

on H^2 by Douglas, Shapiro, and Shields [51] showed that the non-cyclic vectors, *i.e.*, those f for which

$$\bigvee \{ B^n f : n = 0, 1, 2, \dots \} \neq H^2,$$

are precisely those functions Tumarkin identified as representable as in eq.(1.0.4) in the Walsh problem when the poles were too sparse, that is, eq.(1.0.5) holds.

Pseudocontinuations turn up in many investigations involving Hardy spaces, partly because one of the basic building blocks of Hardy space functions, the inner functions, always have pseudocontinuations. For example, Arov [15] (see also Douglas and Helton [49]) encountered pseudocontinuations in the study of matrix-valued inner functions. More precisely, the 'Darlington synthesis' problem, originating in electrical engineering, asks, roughly speaking, when a 'partially filled matrix' can be completed to be matrix-valued inner. In the simplest non-trivial case, 2×2 matrices, the question is the following: what conditions on a bounded analytic function A on $\mathbb{D}$ with $|A| \leq 1$, guarantee the existence of bounded analytic functions B, C, and D such that the matrix

$$\begin{pmatrix} A & B \\ C & D \end{pmatrix}$$

is matrix-valued inner (equivalently has unitary boundary values almost everywhere)? We shall give a brief introduction to this topic in Chapter 6, § 6.7.

Pseudocontinuations and the problem of characterizing the cyclic vectors for the backward shift on H^2 also make connections to the theory of infinite order differential equations. In work of Ritt [116], Pólya [107], Valiron [141], and others, the question arose as to whether or not for a given entire function f, there exists a non-trivial function $\phi = \sum_{n=0}^{\infty} b_n z^n$ of exponential type such that

$$\phi(D)f = 0.$$

Here $D = d/dz$ and $\phi(D)$ is the formal differential operator

$$\phi(D) = b_0 I + b_1 D + b_2 D^2 + \cdots.$$

A known theorem in this area (Proposition 6.8.7) says that a necessary and sufficient condition, for a given f, that there should exist a non-trivial ϕ with $\phi(D)f = 0$, is that the sequence of derivatives $f, f', f'', \dots$ do not span the space of entire functions, endowed with the natural topology of uniform convergence on compacta. In Chapter 6, § 6.8, we consider a different type of infinite order differential equation problem. Here we are given an $f = \sum_{n=0}^{\infty} a_n z^n$ taken from the space V of entire functions with finite norm

$$\|f\|^2 = \sum_{n=0}^{\infty} (n!)^2 |a_n|^2,$$

and ϕ from the class of bounded analytic functions on $\mathbb{D}$, and ask a similar question: when, for given $f \in V$, does there exist a non-trivial bounded analytic function ϕ such that $\phi(D)f = 0$? The answer, that such a ϕ exists if and only if the sequence of derivatives $f, f', f'', \ldots$ does not span V, depends on relating V with the Hardy space H^2 via the unitary operator $U : H^2 \to V$

$$U \left(\sum_{n=0}^{\infty} a_n z^n \right) = \sum_{n=0}^{\infty} \frac{a_n}{n!} z^n.$$

One can show that $UB = DU$, where B is, as above, the backward shift operator

$$Bg = \frac{g - g(0)}{z}$$

on H^2. From here one can apply the Douglas-Shapiro-Shields result, which uses pseudocontinuations to characterize the cyclic vectors for B, to determine which $f \in V$ are the solutions to some non-trivial homogeneous infinite order differential equation $\phi(D)f = 0$ (Theorem 6.8.16). For example, the entire function

$$f = \frac{e^z - 1}{z}$$

belongs to V and is the image of

$$g = \frac{1}{z} \log\left(\frac{1}{1 - z}\right)$$

under U. Since g is a cyclic vector for B on H^2 (due to the isolated winding point at $z = 1$), then, via the identity $UB = DU$, the sequence of derivatives $f, f', f'', \ldots$ span V. Equivalently, the only bounded analytic ϕ on the disk for which $\phi(D)f = 0$ is $\phi \equiv 0$.

It is worth mentioning that the work of Ritt and Pólya on ordinary differential equations of infinite order led to a radically new and very powerful method for proving noncontinuability of some lacunary Taylor and Dirichlet series, yielding, among other things, a new proof of the Fabry gap theorem. Moreover, it revealed the intimate connection of that theorem with results concerning interpolation by entire functions. In recent times, this approach has made multivariable generalizations possible. In Chapter 6, § 6.10, we shall give a brief orientation for the one variable case.

In recent years, this whole subject of 'coherence' or 'continuation' as mentioned in the above paragraphs, especially pseudocontinuation, has been to some extent studied in its own right, in much the same spirit as classical analytic continuation, for example, gap theorems. Gap series

$$\sum_{n=0}^{\infty} a_n z^{\lambda_n}$$

historically have played an important role in creating examples of Taylor series which converge on the open unit disk yet not have an analytic continuation across any arc of the unit circle. Large classes of non-continuable Taylor series were discovered by Hadamard [65] in 1892 and Fabry [55] in 1898. In 1967, one of the present authors [131] began investigating whether or not gap series could be used to create examples of Taylor series which do not have pseudocontinuations across any arc of the unit circle and proved that the gap series

$$\sum_{n=0}^{\infty} 2^{-n} z^{2^n}$$

(which is continuous on $\mathbb{D}^-$ and has a radius of convergence equal to one) is such an example (see Theorem 6.9.7). In Chapter 6, § 6.9, we also point out some remarkable gap series results of Aleksandrov [**7, 8**] which make fascinating connections to number theory. For example, Aleksandrov is able to show that the gap series

$$\sum_{n=1}^{\infty} \frac{1}{n} z^{n^2}$$

does not have a pseudocontinuation across any arc of the circle. However, owing to difficulties of an arithmetic nature, it is unknown whether or not the gap series

$$\sum_{n=1}^{\infty} \frac{1}{n} z^{n^3}$$

has the same property.

In Chapter 7, inspired by work of Bochner and Bohnenblust [**24**], we examine the coherence properties of the two functions

$$f_A(z) = \sum_{n=0}^{\infty} A(n) z^n, \quad |z| < 1$$

$$F_A(z) = -\sum_{n=1}^{\infty} \frac{A(-n)}{z^n}, \quad |z| > 1,$$

where $\{A(n) : n \in \mathbb{Z}\}$ is an 'almost periodic sequence' (see Definition 7.2.8). When $\{c_m\}$ is an absolutely summable sequence, $\{e^{i\theta_m}\}$ is a dense sequence in $\mathbb{T}$, and

$$A(n) = \sum_{m=1}^{\infty} c_m e^{i\theta_m n}, \quad n \in \mathbb{Z},$$

(certainly an almost periodic sequence) then a calculation shows that

$$f_A(z) = \sum_{m=1}^{\infty} \frac{c_m}{1 - e^{i\theta_m} z}, \quad |z| < 1,$$

and F_A is given by the same formula but with $|z| > 1$. Thus, this type of coherence, defined by relating the two component functions f_A and F_A, can be thought of as a generalization of the coherence properties of the Poincaré example (see Chapter 3 and Chapter 7), although it is not clear whether or not f_A and F_A are pseudo-continuations of each other. The main theorem here is the compatibility of the continuation $f_A \to F_A$ with analytic continuation (see Theorem 7.3.1). That is to say, if f_A has an analytic continuation across a boundary point $e^{i\theta}$, then this continuation can be none other than F_A. An analogous result holds when the functions f_A and F_A are replaced by the Laplace transforms

$$f_\phi(z) = \int_0^{\infty} \phi(t) e^{-tz} dt, \quad z = x + iy, \quad x > 0$$

$$F_\phi(z) = -\int_{-\infty}^0 \phi(t) e^{-tz} dt, \quad z = x + iy, \quad x < 0,$$

where ϕ is a Bohr 'almost periodic function' on the real line $\mathbb{R}$, that is to say, the continuation $f_\phi \to F_\phi$ is compatible with analytic continuation.

In Chapter 8, we present a new result, a 'continuation' based on formal multiplication of trigonometric series. In the H^2 setting, we know that $f \in H^2$ is non-cyclic for the backward shift operator (or equivalently can be approximated by rational functions in the Tumarkin sense) if and only if there are bounded analytic functions F and G on the extended exterior disk $\mathbb{D}_e$ such that the non-tangential limits of F/G match the non-tangential limits of f almost everywhere. In terms of boundary values, we say

$$f(e^{i\theta}) = \frac{F}{G}(e^{i\theta}) \ \text{a.e.}$$

or equivalently, the Fourier series of Gf is that of F (considering everything as functions on the circle). In a more general setting (where f is not necessarily in H^2 and F and G are not necessarily bounded), we want to think of an analytic function

$$f = a_0 + a_1 z + a_2 z^2 + \cdots$$

on $\mathbb{D}$ being 'continued' to a meromorphic function F/G on $\mathbb{D}_e$, where

$$F = A_0 + \frac{A_1}{z} + \frac{A_2}{z^2} + \cdots \quad \text{and} \quad G = B_0 + \frac{B_1}{z} + \frac{B_2}{z^2} + \cdots$$

if the *formal* Laurent series of Gf is that of F, that is to say,

$$\left\{ B_0 + \frac{B_1}{z} + \frac{B_2}{z^2} + \cdots \right\}\left\{ a_0 + a_1 z + a_2 z^2 + \cdots \right\} = A_0 + \frac{A_1}{z} + \frac{A_2}{z^2} + \cdots.$$

In the H^2 setting, the above would be equating two trigonometric series to say $Gf = F$ almost everywhere on the circle. We will prove an 'Abelian' type theorem for this trigonometric series continuation making it compatible with ordinary analytic continuation (Theorem 8.2.1). This new type of continuation is, in itself, a concept of independent interest and allows us to identify new classes of cyclic vectors for the backward shift operator on the Bergman and Dirichlet spaces.

We also relate this continuation to the spectral properties of the operator $T := B|\mathcal{M}$, where $\mathcal{M}$ is a non-trivial backward shift invariant (B-invariant) subspace of some reasonable Banach space of analytic functions on $\mathbb{D}$. Problems of this sort were first considered by Moeller [**100**] who characterized the spectrum of T in the Hardy space setting. For $|\lambda| < 1$, it is routine to show that $(I - \lambda T)^{-1}$ exists and moreover,

$$(I - \lambda T)^{-1}f = \frac{zf - \lambda f(\lambda)}{z - \lambda}$$

for every $f \in \mathcal{M}$. When $|\lambda| > 1$ and $(I - \lambda T)^{-1}$ exists, then

$$(I - \lambda T)^{-1}f = \frac{zf - \lambda c_\lambda(f)}{z - \lambda}$$

for some constant $c_\lambda(f)$. Furthermore, whenever the portion of the spectrum of T which lies inside the unit disk is discrete, the function $\lambda \to c_\lambda(f)$ is meromorphic on $\mathbb{D}_e$ and is identical to our new continuation, via formal multiplication of trigonometric series, mentioned above. Many useful results, both about the spectrum of T and the properties of our continuation, can be obtained with this insight.

The subject of generalized analytic continuation began with studying problems in rational approximation and this book comes full circle with that idea. In Chapter 8, § 8.7, we also revisit the Walsh - Tumarkin problem with the Hardy space H^2

replaced with a more general Banach space of analytic functions on $\mathbb{D}$. The generalized Walsh problem was taken up by Gribov and Nikol'skiĭ [**64**]. The generalized Tumarkin problem, originally taken up in a special setting by one of the current authors [**129**], will involve both the spectral properties of the operator $T = B|\mathcal{M}$ as well as our new (formal multiplication of series) continuation.

All the continuations mentioned above (Borel-Walsh, Gončar, pseudocontinuation, continuation via almost periodic sequences, continuation by formal multiplication of series) have corresponding 'Abelian' theorems which places them in compatibility with analytic continuation in much the same way as the classical Abelian theorems of divergent series place the summability methods of Euler, Borel, Abel, and Cesàro, in compatibility with ordinary (Cauchy) convergence. What is so far lacking in the subject of generalized analytic continuation are the 'Tauberian' theorems which compare the various types of continuations. In Chapter 9 we end this book by opening up an investigation of these 'Tauberian' theorems and invite the reader to complete the puzzle. The subject bristles with unsolved problems, some of which shall be enunciated in the following pages.

CHAPTER 2

Notation and Preliminaries

A full listing of the notation used throughout this book can be found in the 'List of Symbols' towards the end of the book.

2.1. Sets

- $\mathbb{C}_\infty = \mathbb{C} \cup \{\infty\}$, the extended complex plane
- $\mathbb{D} = \{z \in \mathbb{C} : |z| < 1\}$, the unit disk.
- $\mathbb{T} = \{z \in \mathbb{C} : |z| = 1\}$, the unit circle.
- $\mathbb{D}_e = \{z \in \mathbb{C}_\infty : 1 < |z| \le \infty\}$, the extended exterior disk.
- A^-, the closure of A.
- A^o, the interior of A.
- A^c, the complement of A.
- If J is a set in some topological vector space, $\bigvee J$ is the closed linear span of the elements of J.
- If J is a set in some topological vector space X with dual X^*, then

$$J^\perp := \{\ell \in X^* : \ell(x) = 0 \ \forall \, x \in A\}$$

  is the annihilator of A.

2.2. Function Spaces

For an open set $\Omega \subset \mathbb{C}_\infty$, let $\mathfrak{H}(\Omega)$ denote the set of single-valued analytic functions on Ω endowed with the metrizable topology of uniform convergence on compact subsets of Ω. Let $\mathfrak{M}(\Omega)$ denote the meromorphic functions on Ω. For a compact subset $K \subset \mathbb{C}$, let $C(K)$ denote the (complex-valued) continuous functions on K endowed with the usual sup-norm topology.

For an analytic function on the unit disk $\mathbb{D}$, we assume the reader is familiar with the notion of a non-tangential limit. Two important limit theorems which will be used many times in this book (and often without much fanfare) will be the classical results of Fatou [56] and Lusin and Privalov [110].

THEOREM 2.2.1 (Fatou). *If $f \in \mathfrak{H}(\mathbb{D})$ and bounded, then f has a finite non-tangential limit at almost every point of $\mathbb{T}$.*

THEOREM 2.2.2 (Lusin and Privalov). *If $f \in \mathfrak{M}(\mathbb{D})$ has non-tangential limits equal to zero on some set of positive measure in $\mathbb{T}$, then f must be identically zero on $\mathbb{D}$.*

These results are covered quite thoroughly in the books of Collingwood and Lohwater [36], Koosis [91], and Privalov [110].

The Hardy space: For $0 < p < \infty$, let H^p (the 'Hardy space') denote the space of $f \in \mathfrak{H}(\mathbb{D})$ for which the quantities

$$M_p(r; f) := \left\{ \int_0^{2\pi} |f(re^{it})|^p \frac{dt}{2\pi} \right\}^{1/p}$$

remain bounded as $r \to 1^-$. This definition can be extended to $p = \infty$ if we define H^∞ to be the bounded analytic functions on $\mathbb{D}$.

The function $r \to M_p(r; f)$ is increasing on $[0, 1)$ and the quantity

$$\|f\|_{H^p} := \lim_{r \to 1^-} M_p(r; f)$$

defines a norm on H^p when $1 \leq p < \infty$. When $0 < p < 1$,

$$\text{dist}(f, g) := \|f - g\|_{H^p}^p$$

defines a metric on H^p. The inequality

$$|f(z)| \leq c_p \|f\|_{H^p} \frac{1}{(1 - |z|)^{1/p}}, \quad z \in \mathbb{D},$$

can be used to show that H^p $(1 \leq p < \infty)$ is a Banach space while H^p $(0 < p < 1)$ is an F-space. For $p = 2$, note that H^2 is a Hilbert space with inner product

$$(2.2.3) \qquad\qquad \langle f, g \rangle = \sum_{n=0}^{\infty} a_n \overline{b_n},$$

where $\{a_n\}$ are the Taylor coefficients (about the origin) of f and $\{b_n\}$ are those of g.

Some of the important properties of H^p spaces that will be used in these notes are the following: given an $f \in H^p$, there is a subset of Lebesgue measure zero in $\mathbb{T}$ off which f has finite non-tangential limits. We denote this limit function by $f(e^{it})$ whenever it exists and note that this boundary function has the important property that it belongs to $L^p := L^p(\mathbb{T}, d\theta)$ and

$$f(re^{it}) \to f(e^{it}) \quad \text{as} \quad r \to 1^-,$$

both almost everywhere as well as in the L^p metric. One can use this to show that the inner product in eq.(2.2.3) can be written as

$$\langle f, g \rangle = \int_0^{2\pi} f(e^{i\theta}) \overline{g(e^{i\theta})} \frac{d\theta}{2\pi}.$$

Every $f \in H^p$ has an associated boundary function which belongs to L^p. We denote this set of boundary functions by

$$H^p(\mathbb{T}) := \{ f \in L^p : f(e^{i\theta}) = \lim_{r \to 1^-} f(re^{i\theta}) \text{ a.e. for some } f \in H^p \}.$$

Turning this problem around, one can ask: when does a given $f \in L^p$ belong to $H^p(\mathbb{T})$? At least for $p \geq 1$, there is an answer given by a theorem of F. and M. Riesz.

THEOREM 2.2.4 (F. and M. Riesz). *For $p \geq 1$, a given $f \in L^p$ belongs to $H^p(\mathbb{T})$ if and only if the Fourier coefficients*

$$\int_0^{2\pi} e^{-int} f(e^{it}) \frac{dt}{2\pi}$$

vanish for all $n < 0$.

Perhaps the most useful fact about H^p functions is that every $f \in H^p$ can be factored as

$$f = O_f I_f,$$

where O_f, the 'outer' factor, is characterized by the property that O_f belongs to H^p and

$$\log|O_f(0)| = \int_0^{2\pi} \log|O_f(e^{it})|\frac{dt}{2\pi}.$$

Every H^p outer function can be written as

(2.2.5) $$F(z) = e^{i\gamma} \exp\left(\frac{1}{2\pi}\int_0^{2\pi}\frac{e^{it}+z}{e^{it}-z}\log\psi(t)dt\right),$$

where γ is a real number, $\psi \geq 0$, $\log\psi \in L^1$, and $\psi \in L^p$. Note that F has no zeros in the open unit disk and $|F(e^{i\theta})| = \psi(\theta)$ almost everywhere. Moreover, every such F as in eq.(2.2.5) belongs to H^p and is outer. The 'inner' factor, I_f, is characterized by the property that I_f is a bounded analytic function on $\mathbb{D}$ whose boundary values satisfy $|I_f(e^{it})| = 1$ for almost every t. Furthermore, the inner factor I_f can be factored further as

$$I_f = bs_\mu,$$

where b is a 'Blaschke product'

$$b(z) = z^m \prod_{n=1}^{\infty}\frac{|a_n|}{a_n}\frac{a_n - z}{1 - \overline{a_n}z}$$

whose zeros at $z = 0$ as well as $\{a_n\} \subset \mathbb{D}\backslash\{0\}$ (repeated according to multiplicity) satisfy the 'Blaschke condition'

(2.2.6) $$\sum_{n=1}^{\infty}(1 - |a_n|) < \infty,$$

and s_μ is the (zero free) 'singular inner factor'

$$s_\mu(z) = \exp\left(-\int_0^{2\pi}\frac{e^{it}+z}{e^{it}-z}d\mu(e^{it})\right)$$

with positive finite singular measure (with respect to Lebesgue measure) μ on $\mathbb{T}$.

Functions of bounded type: An important related space of functions is the meromorphic functions of 'bounded type'

$$\mathfrak{N}(\mathbb{D}) := \{f = \frac{f_1}{f_2} : f_1, f_2 \in H^\infty(\mathbb{D})\}.$$

From the theorems of Fatou (Theorem 2.2.1) and Lusin-Privalov (Theorem 2.2.2), functions in this class have well-defined non-tangential limits almost everywhere on the circle. Two important subclasses of $\mathfrak{N}(\mathbb{D})$ are $N := \mathfrak{N}(\mathbb{D}) \cap \mathfrak{H}(\mathbb{D})$ and the 'Smirnov class' N^+. Every $f \in N$ can be factored as

$$f = b\frac{S_1}{S_2}F,$$

where b is a Blaschke product, S_1, S_2 are singular inner functions, and F is an outer function belonging to N [**52**, p. 25]. Moreover, every such product belongs to N.

The Smirnov class N^+ is the class of functions f as above for which $S_2 \equiv 1$. Two important fact here are (i)

$$H^p \subset N^+ \subset N \subset \mathfrak{N}(\mathbb{D}) \quad \text{for all } p > 0,$$

and (ii) if $f \in N^+$ and the boundary function $f(e^{i\theta})$ belongs to L^p, then $f \in H^p$ (This is false if f is only assumed to be in N). Further information about the basic properties of the H^p spaces can be found in the books of Duren [52], Garnett [60], and Koosis [91].

The Dirichlet space: For an $f \in \mathfrak{H}(\mathbb{D})$ the 'Dirichlet integral' of f is defined to be

$$D(f) = \frac{1}{\pi} \int_{\mathbb{D}} |f'|^2 dA,$$

where dA is two dimensional area measure in the plane. The Dirichlet integral is the area of $f(\mathbb{D})$ (counting multiplicities) and if $f = \sum a_n z^n$, one can compute in polar coordinates to show that

$$D(f) = \sum_{n=0}^{\infty} n |a_n|^2.$$

The 'Dirichlet space' $\mathcal{D}$ is the space of $f \in \mathfrak{H}(\mathbb{D})$ with finite Dirichlet integral $D(f)$. Clearly $\mathcal{D} \subset H^2$ (see eq.(2.2.3)) and thus every $f \in \mathcal{D}$ has boundary values almost everywhere on $\mathbb{T}$. In fact, by a theorem of Beurling [19], Dirichlet functions have non-tangential limits on quite a bit more of the circle. One can define a norm on $\mathcal{D}$ by

$$\|f\|_{\mathcal{D}}^2 := \|f\|_{H^2}^2 + D(f)$$

which turns out to be equal to

$$\sum_{n=0}^{\infty} (n+1)|a_n|^2 = \int_{\mathbb{D}} |(zf)'(z)|^2 dA(z).$$

By means of the estimate,

$$|f(z)|^2 \le \|f\|_{\mathcal{D}}^2 \log \frac{1}{1 - |z|}, \quad z \in \mathbb{D},$$

it follows that $\mathcal{D}$ is Banach space of analytic functions on $\mathbb{D}$. The interested reader can refer to an expository paper of Brown and Shields [31] for the basic properties of the Dirichlet space as well as some related function spaces (see below).

The Bergman space: The 'Bergman space' L_a^2 is the space of $f \in \mathfrak{H}(\mathbb{D})$ for which

$$\|f\|_{L_a^2}^2 := \frac{1}{\pi} \int_{\mathbb{D}} |f(z)|^2 dA$$

is finite. Again, integrating in polar coordinates, one can show that if $f = \sum a_n z^n$, then

$$\|f\|_{L_a^2}^2 = \sum_{n=0}^{\infty} \frac{1}{n+1} |a_n|^2.$$

Using the estimate,

$$|f(z)| \le c \|f\|_{L_a^2} \frac{1}{1 - |z|}, \quad z \in \mathbb{D},$$

it follows that the Bergman space is a Banach space of analytic functions on $\mathbb{D}$. More about the basic properties of Bergman spaces can be found in [46, 71].

The Bergman space is a much larger space of functions than the Hardy space and unlike the Hardy space, Bergman space functions need not have radial limits almost everywhere on $\mathbb{T}$. Indeed, by [**52**, p. 86] there is an $f \in \mathfrak{H}(\mathbb{D})$ which fails to have radial limits on any set of positive measure yet satisfies $|f(z)| \leq (1 - |z|)^{-1/4}$. Such a function satisfying this estimate will belong to the Bergman space. Other such pathological examples can be found in [**85**]. Furthermore, unlike the Hardy space, where the zero sets are precisely those sequences which satisfy the Blaschke condition in eq.(2.2.6), the structure of the zero sets of the Bergman space is more delicate [**79, 123, 124**].

Although the Bergman and Dirichlet spaces are Hilbert spaces and as such are canonically identified with their duals, the duality arguments used later on become easier if we identify the dual of $\mathcal{D}$ with L_a^2 (and vice versa) by means of the 'Cauchy duality'

$$\langle f, g \rangle := \lim_{r \to 1^-} \int_0^{2\pi} f(re^{i\theta})\overline{g}(e^{i\theta})\frac{d\theta}{2\pi}, \quad f \in L_a^2, g \in \mathcal{D}.$$

The r is needed here since, as discussed earlier, Bergman functions need not have well defined boundary functions. If $\{a_n\}$ and $\{b_n\}$ are the Taylor coefficients of f and g respectively, an integral computation yields

$$\langle f, g \rangle = \sum_{n=0}^{\infty} a_n \overline{b_n}.$$

The Dirichlet type spaces: For $\alpha \in \mathbb{R}$, we say an $f = \sum a_n z^n \in \mathfrak{H}(\mathbb{D})$ belongs to the 'Dirichlet type' space D_α if the quantity

$$\|f\|_\alpha^2 = \sum_{n=0}^{\infty} |a_n|^2 (n + 1)^\alpha$$

is finite.

Notice that when $\alpha = -1, 0, 1$, one obtains the Bergman, Hardy, and Dirichlet spaces respectively. When $\alpha < 0$, one can show that

$$\|f\|_\alpha^2 \asymp \int_{\mathbb{D}} |f|^2 (1 - |z|^2)^{-1-\alpha} dA$$

and so D_α is a weighted Bergman space. When $\alpha > 0$, one checks that $f \in D_\alpha$ if and only if $f' \in D_{\alpha-2}$ and so D_α is a space of functions which are 'smooth' up to the boundary. For $\alpha > 1$, not only do the functions from D_α have continuous extensions to $\mathbb{D}^-$, they are Hölder continuous on the circle. Moreover, in this case, D_α is an algebra.

For $\alpha \in \mathbb{R}$, let $h_\alpha : [0, 1) \to \mathbb{R}_+$, be the following function

$$h_\alpha(t) := \begin{cases} (1 - t)^{-(1-\alpha)/2} & \alpha < 1, \\ \sqrt{\log \frac{1}{1-t}} & \alpha = 1, \\ 1 & \alpha > 1 \end{cases}$$

It is known [**31**, Prop. 10], that for any $g \in D_\alpha$,

(2.2.7) $$|g(z)| = C_\alpha \|g\|_{D_\alpha} h_\alpha(|z|), \quad z \in \mathbb{D},$$

and so it follows that D_α is a Banach space of analytic functions on $\mathbb{D}$.

Again, even though D_α is a Hilbert space and hence its own dual, it will be more convenient to identify the dual of D_α with $D_{-\alpha}$ by means of the 'Cauchy duality'

$$\langle f, g \rangle = \sum_{n=0}^{\infty} a_n \overline{b_n} = \lim_{r \to 1^-} \int_0^{2\pi} f(re^{i\theta}) \overline{g}(re^{i\theta}) \frac{d\theta}{2\pi},$$

where $\{a_n\}$ and $\{b_n\}$ are the Taylor coefficients of f and g respectively. We refer the reader to [**31**] for a thorough study of the basic properties of these spaces.

Spaces of entire functions: Let E denote the space of entire functions. With the metrizable topology of uniform convergence on compacta, E is a topological vector space whose dual can be identified with the space X of entire functions of 'exponential type', $i.e.$, those entire $g = \sum b_n z^n$ which satisfy

$$n! |b_n| = O(A_g^n)$$

for some constant A_g. Equivalently, $g \in X$ if and only if g satisfies

$$|g(z)| \leq Ce^{B_g |z|}.$$

The dual pairing between E and X can be written as

(2.2.8) $$[f, g] := \sum_{n=0}^{\infty} n! a_n b_n,$$

where $f = \sum a_n z^n \in E$ and $g = \sum b_n z^n \in X$. Note that since f is entire, then

$$\lim_{n \to \infty} \sqrt[n]{|a_n|} = 0$$

and so the above series which defines $[f, g]$ converges. When considering the space X throughout these notes, we will be assuming that X is endowed with the weak-star topology it inherits by being the dual of E. The dual of X can then be identified with E via the above pairing. A nice reference for this material is a book of Levin [**95**].

CHAPTER 3

The Poincaré example

In this chapter we discuss in greater detail the Poincaré example,

$$f = \sum_{n=1}^{\infty} \frac{c_n}{1 - e^{-i\theta_n} z},$$

where $\{e^{i\theta_n}\}$ is a sequence of distinct points which are dense set in $\mathbb{T}$ and $\{c_n\}$ is an absolutely summable sequence of non-zero complex numbers. This example is important for several reasons. For one, it is an example of an analytic function on $\mathbb{C}_\infty \setminus \mathbb{T}$ which arises from a single formula but for which the unit circle is a *coupure* (cut) or natural boundary for both $f|\mathbb{D}$ and $f|\mathbb{D}_e$. We will provide a proof of this shortly. Secondly, although the component functions $f|\mathbb{D}$ and $f|\mathbb{D}_e$ do not have analytic continuations across any point of $\mathbb{T}$, they can be thought of as 'continuations' of each other since they have non-tangential boundary values which are equal almost everywhere. As mentioned in the Overview, this 'matching non-tangential boundary values' continuation is known as pseudocontinuation and appears in numerous places. A detailed discussion of pseudocontinuations will take place in Chapter 6.

3.1. Poincaré's result

THEOREM 3.1.1 (Poincaré [**106**]). *Let L be a smooth closed curve which bounds a convex set in the plane, $\{z_n\}$ be a sequence of distinct points in L which are dense in L, and $\{c_n\}$ be an absolutely summable sequence of non-zero complex numbers. If f is the function*

$$f(z) := \sum_{n=1}^{\infty} \frac{c_n}{z - z_n}, \quad z \notin L,$$

then $f|int(L)$ does not have an analytic continuation across any point of L.

REMARK 3.1.2. As mentioned in our Overview, the restriction that L bounds a convex set was removed by Denjoy [**43**]. Although we can prove a somewhat stronger result (which we will later), we include a variation of Poincaré's original proof.

PROOF OF THEOREM 3.1.1. First, having chosen one of the z_k, one can easily verify using the hypothesis that one can pick w in the interior of L such that the closed disk centered at w with radius $|w - z_k| := R$ meets L only at z_k. We fix w, which, by translating the figure, we may assume to be the origin.

The function f has a Taylor series about the origin whose Taylor coefficients are equal to

$$B_q = -\sum_{n=0}^{\infty} \frac{c_n}{z_n^{q+1}}, \quad q = 0, 1, 2, \ldots$$

We will show this Taylor series has radius of convergence precisely R, implying f cannot have an analytic continuation across z_k. To do this, it suffices to show

$$R^q B_q \nrightarrow 0 \text{ as } q \to \infty.$$

Notice

$$(3.1.3) \qquad R^q B_q = -\frac{c_k R^q}{z_k^{q+1}} - \sum_{n \neq k} \frac{c_n R^q}{z_n^{q+1}}$$

and

$$\left| -\frac{c_k R^q}{z_k^{q+1}} \right| = \frac{|c_k|}{R}.$$

We finish the proof by showing that the second term in eq.(3.1.3) goes to zero as $q \to \infty$. To do this, note that for $n \neq k$, $|R/z_n| < 1$ and apply the Lebesgue dominated convergence theorem. $\qquad \square$

REMARK 3.1.4.   (1) A similar proof shows $f|\text{ext}(L)$ does not have an analytic continuation across any point of L.
   (2) The proof given here is essentially Poincaré's original argument except that he needed to take a longer path to prove $R^q B_q \nrightarrow 0$ since his paper was published in 1883 before the Lebesgue dominated convergence theorem.

3.2. Matching non-tangential boundary values

Let the curve L be the unit circle $\mathbb{T}$ in the previous Poincaré result. Although the component functions $f|\mathbb{D}$ and $f|\mathbb{D}_e$ do not have analytic continuations across any point of the circle, they do display some properties of 'coherence'. The first form of coherence, though rather trivial, is the following.

PROPOSITION 3.2.1. *If $f|\mathbb{D} \equiv 0$, then $f|\mathbb{D}_e \equiv 0$.*

The proof follows from this more general result which can be obtained from the dominated convergence theorem.

PROPOSITION 3.2.2. *Let $\{e^{i\theta_n}\}$ be a sequence of distinct points in $\mathbb{T}$ and $\{c_n\}$ be an absolutely summable sequence of complex numbers. The function*

$$(3.2.3) \qquad f(z) := \sum_{n=1}^{\infty} \frac{c_n}{1 - ze^{-i\theta_n}}, \quad z \in \mathbb{D},$$

has the following property:

$$(3.2.4) \qquad \lim_{r \to 1^-} (1 - r)f(re^{i\theta_m}) = c_m, \quad m = 1, 2, \ldots$$

This type of coherence ($f|\mathbb{D} \equiv 0$ implies $f|\mathbb{D}_e \equiv 0$) is certainly not the strongest result we get here. We now discuss the much stronger coherence between the component functions $f|\mathbb{D}$ and $f|\mathbb{D}_e$ via matching non-tangential boundary values. Before we can show these non-tangential boundary values *match*, we must first establish they actually *exist*. Though this next lemma is quite standard in survey books on H^p spaces [**52**, p. 39] [**60**, p. 114], we include a proof.

DEFINITION 3.2.5. Let μ be a bounded complex Radon measure on $\mathbb{T}$ and let f_μ be the analytic function defined on $\mathbb{C}_\infty \backslash \mathbb{T}$ by

$$f_\mu(z) := \int \frac{1}{1 - e^{-it}z} d\mu(e^{it}).$$

The function f_μ is often called the 'Cauchy transform' of μ.

THEOREM 3.2.6 (Smirnov [**135**]). *If μ is a bounded complex Radon measure on $\mathbb{T}$, then*

$$f_\mu \in \bigcap_{0 < p < 1} H^p.$$

In fact, $\|f_\mu\|^p \leq A_p \|\mu\|^p$, where $\|\mu\|$ is the total variation norm of μ.

Writing μ as a (complex) linear combination of positive measures and noting that for a positive measure μ,

$$\Re(f_\mu(z)) = \int \frac{1 - \Re(e^{-it}z)}{|1 - e^{-it}z|^2} d\mu(e^{it}) > 0, \quad z \in \mathbb{D},$$

the result follows immediately from the following lemma, which is also quite standard [**60**, p. 114].

LEMMA 3.2.7. *Let $F \in \mathfrak{H}(\mathbb{D})$ with $\Re F > 0$. Then for all $0 < r < 1$ and $0 < p < 1$,*

$$\int_0^{2\pi} |F(re^{i\theta})|^p d\theta \leq A_p |F(0)|^p.$$

PROOF. Since $\Re F > 0$, then $F = |F|e^{i\phi}$, where $-\pi/2 < \phi < \pi/2$. Also observe that since F has no zeros in the disk, F^p (the branch which has positive real part at the origin) is also analytic on $\mathbb{D}$ and

$$F^p = |F|^p \left(\cos(p\phi) + i\sin(p\phi) \right).$$

For $0 < p < 1$,

$$\Re(F^p) = |F|^p \cos(p\phi) \geq |F|^p \cos(p\pi/2).$$

From this we conclude

$$\int_0^{2\pi} |F(re^{i\theta})|^p d\theta \leq A_p \int_0^{2\pi} \Re(F^p(re^{i\theta})) d\theta = A_p \Re(F^p(0)).$$

The last equality follows from the fact that $\Re(F^p)$ is harmonic. The desired inequality follows from the observation $\Re(F^p(0)) \leq |F(0)|^p$. $\square$

The function f in eq.(3.2.3) is the Cauchy transform of the finite measure

(3.2.8)
$$d\mu = \sum_{n=1}^{\infty} c_n \delta_{e^{i\theta_n}}$$

and so $f|\mathbb{D}$ and $f|\mathbb{D}_e$ belong to H^p (resp. $H^p(\mathbb{D}_e)$ [1]) giving them, by standard H^p theory (see Chapter 2), finite non-tangential boundary values almost everywhere. To show these boundary values are indeed equal for almost every $e^{i\theta}$, we make a few observations.

First, we know that for any bounded complex Radon measure μ on $\mathbb{T}$, the limits of $f_\mu(re^{i\theta})$ and $f_\mu(e^{i\theta}/r)$ exist as $r \to 1^-$. Thus, to show that the non-tangential limits are equal for almost every $e^{i\theta}$, it suffices to prove that when μ is singular with respect to Lebesgue measure on the circle, as is the case with the measure in eq.(3.2.8),

$$f_\mu(re^{i\theta}) - f_\mu(e^{i\theta}/r) \to 0 \quad \text{as } r \to 1^-$$

for almost every $e^{i\theta}$. To this end, note that

$$f_\mu(re^{i\theta}) - f_\mu(e^{i\theta}/r) = \int P_{re^{i\theta}}(e^{it})d\mu(e^{it}),$$

where

$$P_{re^{i\theta}}(e^{it}) = \frac{1-r^2}{|e^{it}-re^{i\theta}|^2}$$

is the usual Poisson kernel. Consider the Lebesgue decomposition of $d\mu$ with respect to arc length measure $d\theta$ on $\mathbb{T}$, and let $M \in L^1(\mathbb{T})$ denote the Radon-Nikodym derivative of its absolutely continuous part with respect to $d\theta$ (observe that for the above purely atomic measure as in eq.(3.2.8), $M = 0$). We shall use the following theorem of Fatou [77, p. 34] to deduce that for the measure in eq.(3.2.8),

$$\int P_{re^{i\theta}}(e^{it})d\mu(e^{it}) \to 0$$

for almost every $e^{i\theta}$ as $r \to 1^-$.

THEOREM 3.2.9 (Fatou). *If μ is a finite measure on $\mathbb{T}$, then*

$$\lim_{r\to 1^-} \int P_{re^{i\theta}}(e^{it})d\mu(e^{it}) = 2\pi M(e^{i\theta})$$

for almost every $e^{i\theta}$.

REMARK 3.2.10. A variant of Poincaré's example, involving almost periodic functions, will be taken up in Chapter 7.

---

[1] $H^p(\mathbb{D}_e)$, the Hardy space of the exterior disk, is the set of $F(z) = f(1/z)$, where $f \in H^p$.

CHAPTER 4

Borel's ideas and their later development

As mentioned in our Overview, the early work of Borel pointed towards the coherence of the component functions $f|\Omega_j$, where

$$f(z) = \sum_{n=1}^{\infty} \frac{c_n}{z - z_n}$$

and $\{\Omega_j : j = 1, 2, 3, \ldots\}$ are the connected components of $\mathbb{C}_\infty \backslash \{z_n\}^-$. In this chapter, we discuss the coherence properties of the component functions $f|\Omega_j$ when the partial sums of the above 'Borel series' are replaced by a more general type of rational approximation we call 'superconvergence'.

4.1. Superconvergence

Let A be a compact set in the plane, and $\{P_n\}, \{Q_n\}$ be two sequences of (analytic) polynomials with complex coefficients satisfying

$$\deg P_n \leq n, \ \deg Q_n \leq n,$$

$$Q_n^{-1}(\{0\}) \subset A,$$

$$P_n \text{ and } Q_n \text{ have no common roots.}$$

Define

(4.1.1) $$R_n = P_n/Q_n.$$

Thus R_n is a rational function with at most n zeros in the plane and at most n poles in A.

DEFINITION 4.1.2. We say a sequence of rational functions $\{R_n\}$, as in eq.(4.1.1), is 'superconvergent' on a compact subset $K \subset A^c$ if

(4.1.3) $$\max \{ |R_n(z) - R_{n+1}(z)| : z \in K \}^{1/n} \to 0 \text{ as } n \to \infty.$$

PROPOSITION 4.1.4. *For a sequence of rational functions $\{R_n\}$, as in eq.(4.1.1), and a compact set $K \subset A^c$, the following are equivalent.*

(1) *$\{R_n\}$ is superconvergent on K.*
(2) *There is an $f \in C(K) \cap \mathfrak{H}(K^o)$ such that*

(4.1.5) $$\max \{ |f(z) - R_n(z)| : z \in K \}^{1/n} \to 0 \text{ as } n \to \infty.$$

PROOF. If eq.(4.1.3) holds and

$$a_n := \max \{ |R_n(z) - R_{n+1}(z)| : z \in K \},$$

21

then $\sqrt[n]{a_n} \to 0$ and so $\sum a_n < \infty$. Hence

$$f(z) := \lim_{n \to \infty} R_n(z)$$

exists for each $z \in K$ and moreover, $f \in C(K) \cap \mathfrak{H}(K^o)$. Writing

$$b_n := \max \left\{ |f(z) - R_n(z)| : z \in K \right\},$$

and repeatedly using the inequality

$$|f(z) - R_j(z)| \le |R_j(z) - R_{j+1}(z)| + |f(z) - R_{j+1}(z)|,$$

we have

$$b_n \le a_n + a_{n+1} + \cdots$$

It follows now that $\sqrt[n]{b_n} \to 0$.

Conversely, if eq.(4.1.5) holds, $i.e.$, $\sqrt[n]{b_n} \to 0$ for some $f \in C(K) \cap \mathfrak{H}(K^o)$,

$$|R_n(z) - R_{n+1}(z)| \le b_n + b_{n+1}$$

for all $z \in K$ and so $\sqrt[n]{a_n} \to 0$. $\square$

For example, if $\{z_j\}$ is a bounded sequence of distinct points in $\mathbb{C}$ and $\{c_j\}$ is a sequence of non-zero complex numbers which satisfy

$$\lim_{n \to \infty} \sqrt[n]{|c_n|} = 0,$$

then the rational functions defined by

$$R_n(z) := \sum_{j=1}^{n} \frac{c_j}{z - z_j}$$

are superconvergent on every compact subset $K \subset \mathbb{C} \backslash \{z_j\}^-$. The limit function

$$\sum_{j=1}^{\infty} \frac{c_j}{z - z_j}$$

is the function considered in the work of Borel (see the Overview).

DEFINITION 4.1.6. A 'continuum' is a compact connected set in $\mathbb{C}$. A 'non-degenerate continuum' is a continuum which contains more than one point.

We will show in a moment that if $\{R_n\}$ is superconvergent on a non-degenerate continuum $K \subset A^c$, then $\{R_n\}$ is superconvergent on any compact subset of A^c. To be able to do this, we recall two classical results of Bernstein and Faber and refer the reader to Walsh's book [**143**, pp. 77 - 78] for additional details. First, however, we state an elementary conformal mapping result [**99**, Vol. 3, p. 104].

LEMMA 4.1.7. *Let K be a non-degenerate continuum with connected complement K^c and let $G = \mathbb{C}_\infty \backslash K$, the complement of K in the extended complex plane. Then there is a function ϕ which maps G conformally onto a domain of the form $\{|z| > R\}$, for some $R > 0$, and satisfies the conditions*

$$\phi(\infty) = \infty \quad and \quad \lim_{z \to \infty} \phi(z)/z = 1.$$

LEMMA 4.1.8 (Bernstein - Faber lemma). *Let P be a polynomial in the complex variable z of degree no more than n, and K be a non-degenerate continuum with connected complement K^c. Let ϕ be the conformal map of K^c onto $\{|w| > R\}$ as in Lemma 4.1.7. Then,*

$$|P(a)| \leq M(|\phi(a)|/R)^n, \quad \text{for all } a \in K^c,$$

where $M = \max\{|P(z)| : z \in K\}$.

PROOF. The function $g := P/\phi^n$ is analytic on K^c with finite limiting value at infinity and hence is regular at infinity. By the maximum principle,

$$|g(a)| \leq \sup\left\{ |g(z)| : z \in K \right\} \leq \frac{M}{R^n}, \quad a \in K^c,$$

and the result follows. □

THEOREM 4.1.9 (Bernstein - Faber theorem). *The necessary and sufficient condition for a continuous function g on a non-degenerate continuum K with connected complement K^c to have distance d_n from the polynomials of degree n, in the metric of $C(K)$, satisfying*

$$\rho := \varlimsup_{n \to \infty} \sqrt[n]{d_n} < 1,$$

is that there exists an analytic function G on a neighborhood of K with $G|K = g$.

REMARK 4.1.10. The above two results were shown by Bernstein for the case when the continuum K is a line segment. The general case we use here was shown by Faber and later generalized by Walsh (see [**143**, p. 78] for the proper historical references). The size of the domain of holomorphy in terms of ρ and the level sets of the Green's function of K^c can be given explicitly (see [**143**] and [**99**, Vol. 3, p. 114]).

Returning to our problem of superconvergence, we can apply Lemma 4.1.8 to obtain the following useful result.

PROPOSITION 4.1.11. *If $\{R_n\}$, as in eq.(4.1.1), is a superconvergent sequence of rational functions (with poles in A) on a non-degenerate continuum K with connected complement K^c, then $\{R_n\}$ is superconvergent on any compact subset of A^c.*

PROOF. Suppose $\{R_n\}$ is superconvergent on K. There is no loss of generality in assuming the polynomials Q_n are monic (*i.e.*, have leading coefficient equal to one). We now show that

$$\max\left\{ |R_n(z) - R_{n+1}(z)| : z \in L \right\} \leq a_n M^n,$$

where L is a compact subset of A^c, $M = M(A, K, L)$ is a constant depending only on A, K, L, and

$$a_n = \max\{|R_n(z) - R_{n+1}(z)| : z \in K\}.$$

Indeed, on K we have the estimate

(4.1.12) $$|Q_n(z)| \leq C^n,$$

where $C = C(A, K)$ depends only on A and K. One can see this by noting that

$$|Q_n(z)| = |z - a_{n,1}| \cdots |z - a_{n,k_n}|,$$

where $a_{n,j} \in A$ and $k_n := \deg Q_n \leq n$. Thus $|Q_n(z)| \leq (m + a + 1)^n$, where $m = \max\{|z| : z \in K\}$ and $a = \max\{|z| : z \in A\}$.

Thus the denominator of the rational function $R_n - R_{n+1}$, which is $Q_n Q_{n+1}$, is bounded above on K, in absolute value, by C^{2n+1}. Hence the numerator

$$P_n Q_{n+1} - Q_n P_{n+1}$$

is bounded above, in absolute value, on K by $a_n C^{2n+1}$. Using Lemma 4.1.8, this numerator is bounded on L by

$$a_n C^{2n+1} B^{2n+1},$$

where $B = B(K, L)$ is a constant depending only on K and L. Also, for $z \in L$,

$$(4.1.13) \qquad |Q_n(z)| \geq d^{k_n} \geq \left(\frac{d}{1+d}\right)^{k_n} \geq \left(\frac{d}{1+d}\right)^n,$$

where $d = \mathrm{dist}(L, A)$. Thus, $|Q_n(z) Q_{n+1}(z)| \geq s^{2n+1}$ for $z \in L$. Hence,

$$\max \left\{ |R_n(z) - R_{n+1}(z)| : z \in L \right\} \leq a_n \left(\frac{CB}{s} \right)^{2n+1},$$

whose n-th root tends to zero as $n \to \infty$ (since $\sqrt[n]{a_n} \to 0$ by assumption), completing the proof. $\qquad\square$

We can now prove the following basic theorem of Walsh.

THEOREM 4.1.14 (Walsh [143]). *Suppose f is analytic on a connected open set Ω and, on some non-degenerate continuum $K \subset \Omega$ with connected complement K^c, f is the limit of a superconvergent sequence $\{R_n\}$ of rational functions. Then R_n converges, indeed it superconverges, to f on every compact subset L of $\Omega \setminus A$.*

REMARK 4.1.15. The point here is that A is permitted to meet Ω, even to disconnect it, yet the superconvergence 'survives' even if L is separated from K by the poles of the R_n's.

PROOF. By hypothesis, and the preceding discussion,

$$|f(z) - R_n(z)| \leq b_n, \quad \text{for all } z \in K,$$

where $\sqrt[n]{b_n} \to 0$. For all $z \in K$,

$$(4.1.16) \qquad |f(z) Q_n(z) - P_n(z)| \leq b_n |Q_n(z)| \leq b_n E^n,$$

where $E = E(A, K)$ (see eq.(4.1.12) and note that we are assuming Q_n is monic). Moreover, if Γ is a smooth curve in Ω which surrounds both K and L, we have

$$(4.1.17) \qquad |f Q_n - P_n| \leq M^n \quad \text{on } \Gamma,$$

for some constant M depending on f, A, K, L, and Γ (but not on n). Indeed, for $z \in \Gamma$,

$$|f(z) Q_n(z) - P_n(z)| \leq C(f, \Gamma) |Q_n(z)| + |P_n(z)| \leq C(f, \Gamma) C(A, \Gamma)^n + |P_n(z)|$$

by eq.(4.1.12). By Proposition 4.1.11, $\{R_n\}$ superconverges on Γ and so, in particular, $|R_n|$ is uniformly bounded by C_Γ on Γ. Thus for $z \in \Gamma$,

$$|P_n(z)| \leq C_\Gamma |Q_n(z)| \leq C_\Gamma C(A, \Gamma)^n$$

which verifies eq.(4.1.17).

If we now apply the two-constant theorem [75, p. 314] [1] to estimate $fQ_n - P_n$ on L (considered as a subset of the domain bounded by Γ and the boundary of K), we get, using eq.(4.1.16) and eq.(4.1.17),

$$\max\left\{ |f(z)Q_n(z) - P_n(z)| : z \in L \right\} \leq (b_n E^n)^s (M^n)^{1-s}$$

for some $0 < s < 1$ determined only by the geometrical configuration. Using eq.(4.1.13) we get that $|f - R_n|$ is bounded above on L by

$$\frac{1}{s^n}(b_n E^n)^s (M^n)^{1-s}$$

where $s = d/(1+d)$ and $d = \operatorname{dist}(A, L)$. The n-th root of this goes to zero as $n \to \infty$ (owing to the factor b_n), which completes the proof. $\qquad\square$

4.2. Borel series

The theorem just proven has several interesting consequences. For one thing, suppose f is analytic on the interior of some disk K, where it is the limit of a superconvergent sequence of rational functions having poles in K^c. If f has analytic continuations along two paths leading from K to a neighborhood of some point $w \in A^c$ (regardless of the paths - in particular they may intersect A!) then the continued functions coincide at w. Expressed less precisely: the limit of a superconvergent sequence can never display multiple-valuedness. This is interesting from the standpoint of approximation theory, for example, the function 'square root of z' on the interval $[1, 2]$ cannot be the limit of a superconvergent sequence of rational functions with poles in a compact set A disjoint from $[1, 2]$ and the point 0.

In line with the general theme of this book, *i.e.*, generalized analytic continuation, if $\{\Omega_j : j = 1, 2, \ldots\}$ are the connected components of A^c, for a compact set A, and $f \in \mathfrak{H}(A^c)$ is the superconvergent limit of rational functions whose poles are in A, then the component functions $f|\Omega_j$ can be regarded as 'continuations' of each other. The Walsh theorem shows that this 'Borel-Walsh' continuation, via approximation by superconvergent rational functions, is compatible with analytic continuation in the sense that if $f|\Omega_j$ has an analytic continuation along some path to a neighborhood of a point of Ω_k, then this analytic continuation can be none other than $f|\Omega_k$. More precise statements can be made refining the crude analysis above, in line with later work of Walsh and Gončar, and will be presented below (see especially Chapter 9). We also have the following consequence, for which we recall the notion of Borel series from the Overview.

---

[1] The two constant theorem: Let f be analytic and bounded above by M on a domain Ω which is bounded by a finite number of simple closed curves C. Suppose there exists a sub-arc Γ of C (or arcs) such that $\overline{\lim}_{z \to \Gamma} |f(z)| \leq m < M$. Then there is a function $\lambda(z)$ with $0 < \lambda(z) < 1$ such that for each $z \in \Omega$, $\log|f(z)| \leq \lambda(z) \log m + (1 - \lambda(z)) \log M$. Moreover, on compact subsets of Ω, the function λ is bounded away from 0 and 1. Actually, $\lambda(z)$ is the harmonic measure of Γ with respect to z.

DEFINITION 4.2.1. Let $\{z_n\}$ be a bounded sequence of distinct points in $\mathbb{C}$ and $A = \{z_n\}^-$. If $\{c_n\}$ are complex numbers such that

$$(4.2.2) \qquad \sum_{n=1}^{\infty} |c_n| < \infty,$$

let the 'Borel series' be the function defined by

$$\sum_{n=1}^{\infty} \frac{c_n}{z - z_n}.$$

The above series converges uniformly on compact subsets of A^c to an analytic function f which clearly extends meromorphically to a neighborhood of each isolated point of $\{z_n\}$.

In the interesting case where A disconnects the plane, it is not clear what, if any, relations will exist between the restrictions of f to the various components of A^c. Indeed, it was long an open question, raised by Borel, whether one of these restrictions could be zero without all the remaining ones vanishing. In the case where the coefficients $\{c_n\}$ decrease so rapidly that

$$(4.2.3) \qquad \lim_{n \to \infty} \sqrt[n]{|c_n|} = 0,$$

the partial sums of the Borel series are superconvergent, so the earlier theorem of Walsh (Theorem 4.1.14) implies this is impossible. Note that Borel [**27, 28**] actually proved this coherence result in the special case of a Borel series.

If, however, only the summability condition in eq.(4.2.2) is assumed, J. Wolff presented an example where f vanishes only on a single component of A^c. More precisely, he constructed a (not identically zero) Borel series, the sum of which is zero in the unit disk. This very elegant example will be recalled shortly. Actually, Wolff presented two different constructions in consecutive *Comptes Rendus* notes in 1921 [**146**]. In the first construction, the $\{z_n\}$ cluster not only at all points of $\mathbb{T}$ (an obvious necessary condition for such an example), but also at some points in $\mathbb{D}_e$. This aesthetic blemish was removed in Wolff's second construction, based on a simple application of Cauchy's integral theorem. We shall present both constructions, as well as a third based on a recent result of Bonsall [**26**]. Bonsall's method may be compared with Wolff's second construction: Wolff's has the advantage of yielding a constructive algorithm, whereas the former is based on functional analysis and non-constructive. On the other hand, the Bonsall result yields more, and shows that any sequence $\{z_n\}$ clustering non-tangentially at almost every boundary point of $\mathbb{T}$ can be used as the set of 'poles' in a Borel series whose sum in $\mathbb{D}$ is any prescribed function analytic on a neighborhood of $\mathbb{D}^-$. A paper of Brown, Shields, and Zeller [**32**] brought this idea to its complete fruition (see Theorem 4.2.12 below).

The drastic hypothesis in eq.(4.2.3) is sufficient to exclude the above pathological behavior and thus ensuring that the Borel series is a *fonction monogène* in Borel's sense [**27**]. In a 'local' variant, one can replace the condition in eq.(4.2.3) by the weaker assumption of exponential decay, but this is essentially the limit of what is possible. Denjoy [**42, 43**], Beurling [**18, 22**], Leont'eva [**94**] have all constructed (non-trivial) Borel series which sum to zero on a disk for which the coefficients $\{c_n\}$ are just shy of exponential decay. For example, Denjoy [**42**] constructed such a Borel series with

$$|c_n| \le k \exp\left(-n^{1/2-\varepsilon}\right)$$

while Beurling [**18, 22**] constructed one with

$$|c_n| \le \exp\left(-n/(\log n)^2\right)$$

and even closer approaches to exponential decay involving iterated logarithms. We now present the above-mentioned constructions of Wolff.

Wolff's first example: To get us started, we need the following lemma whose proof we leave to the reader.

LEMMA 4.2.4. *Let E be an open set in the plane with $Area(E) < \infty$. Then, there exists a sequence $\{D_j\}$ of mutually disjoint closed disks contained in E such that*

$$\sum_{j=1}^{\infty} Area(D_j) = Area(E).$$

To obtain Wolff's example of a Borel series which vanishes on only one of its components, apply the previous lemma to $E := \{z : 1/2 < |z| < 1\}$. That is, let $\{D_j\}$ be pairwise disjoint disks contained in E and z_j be the center of D_j. Then, by the mean value property for harmonic functions, for any function u continuous in $\mathbb{D}^-$ and harmonic in $\mathbb{D}$, we have

$$u(0) = \frac{1}{\text{Area}(E)} \int_E u \, dA = \frac{1}{\text{Area}(E)} \sum_{j=1}^{\infty} \int_{D_j} u \, dA,$$

(where dA is two-dimensional Lebesgue measure in the plane) since the D_j's cover E except for a set of area measure zero. Again, by the mean value property, this last expression equals

$$\frac{1}{\text{Area}(E)} \sum_{j=1}^{\infty} \text{Area}(D_j) \, u(z_j).$$

Putting this all together,

$$u(0) = \sum_{j=1}^{\infty} b_j u(z_j),$$

where $b_j := \text{Area}(D_j)/\text{Area}(E)$ are positive numbers which sum to one.

In particular, we can apply this to $u(z) = 1/(w - z)$, where w is a point in $\mathbb{D}_e$, and get

$$\frac{1}{w} = \sum_{j=1}^{\infty} \frac{b_j}{w - z_j}$$

which is valid for all $w \in \mathbb{D}_e$. By transposing the left hand term to the right side, we have a Borel series - with poles at zero and the $\{z_j\}$ - whose sum is identically zero on $\mathbb{D}_e$.

Wolff's second example: The second example of a non-trivial Borel series which vanishes on one of its components is contained in the following theorem.

THEOREM 4.2.5 (Wolff [**146**]). *Let G be any Jordan domain, and f be analytic on a neighborhood of G^-. Then, there is a sequence $\{z_n\} \in (G^-)^c$ and an absolutely* *summable sequence of complex numbers $\{c_n\}$ such that*

$$f(z) = \sum_{n=1}^{\infty} \frac{c_n}{z - z_n} \quad \text{for } z \in G^-.$$

It will be clear from the constructive nature of the proof of the above theorem that given a point $a \in (G^-)^c$, one can represent the function $1/(z - a)$ on G as a Borel series above but with $z_n \neq a$ for any n. From here, we obtain a non-trivial Borel series (with poles at the z_n's as well as the point a) summing identically to zero on G.

We will prove Wolff's theorem under the assumption that G is the unit disk $\mathbb{D}$. The general case requires only trivial modifications. We first need to prove the following lemma.

LEMMA 4.2.6. *Let f be analytic for $|z| < c$, and $0 < a < b < c$. Then, given $\varepsilon > 0$ there is a rational function*

$$R(z) = \sum \frac{c_n}{z - z_n},$$

where $\{z_n\}$ is a finite subset of $\{|z| = b\}$, and $\{c_n\}$ are complex numbers with

$$\sum |c_n| \leq cM(f, b),$$

where $M(f, b) := \max\{|f(be^{it})| : 0 \leq t \leq 2\pi\}$, such that

$$|f(z) - R(z)| \leq \varepsilon \quad \text{for all } |z| \leq a.$$

PROOF. By Cauchy's formula, we have for $|z| \leq a$,

$$f(z) = \frac{1}{2\pi i} \int_{|w|=b} \frac{f(w)}{w - z} dw.$$

The above integral can be written as

$$\int_0^{2\pi} \frac{1}{2\pi i} \frac{f(be^{it})}{be^{it} - z} ibe^{it} dt.$$

If we denote the integrand, for fixed z, by $F(t)$, then the Riemann sum

$$S_N := \frac{2\pi}{N} \sum_{k=0}^{N-1} F\left(\frac{2\pi k}{N}\right)$$

approaches the integral as $N \to \infty$, uniformly for $|z| \leq a$. Indeed, the integral differs from S_N by a quantity of magnitude not exceeding $1/N$ times the total variation of the integrand, see [109, p. 49]. This total variation is clearly uniformly bounded for $|z| \leq a$. Choosing for R the sum S_N with N sufficiently large yields the conclusion of the lemma.   □

PROOF OF WOLFF'S THEOREM. By hypothesis, there is a number $c > 1$ such that f is analytic for $|z| < c$. Let L_m denote the circle $\{|z| = b_m\}$, where $\{b_m\}$ is any strictly decreasing sequence of numbers tending to 1, and with $b_1 < c$. Also, there is no loss of generality in assuming that $M(f, b_1) \leq 1$.

For a rational function of the form

(4.2.7) $$R(z) = \sum \frac{c_n}{z - z_n},$$

we will denote the (finite) sum $\sum |c_n|$ by $[R]$. Applying the previous lemma, we can produce a rational function of the form in eq.(4.2.7), with all z_n on L_1 and $[R] \leq cM(f, b_1)$, satisfying $M(f - R, b_2) \leq c/2$ (here we choose $\varepsilon = c/2$).

To make the presentation as clear as possible, let us denote f henceforth as f_1 and the R just constructed as R_1. Thus, denoting $f_2 := f_1 - R_1$, f_2 is analytic inside L_1 and $M(f_2, b_2) \leq c/2$. Also, $[R_1] \leq M(f_1, b_1) \leq c$. Applying the lemma again, we obtain a rational function R_2 with (simple) poles on L_2, satisfying $M(f_3, b_3) \leq c/4$ (where $f_3 := f_2 - R_2$), and

$$[R_2] \leq M(f_2, b_2) \leq c/2.$$

Proceeding inductively in this manner, we obtain sequences $\{f_n\}$ and $\{R_n\}$, where

$$f_{n+1} := f_n - R_n,$$

R_n has all poles on L_n, and

$$M(f_n, b_n) \leq \frac{c}{2^n}, \quad [R_n] \leq \frac{c}{2^{n-1}}.$$

For each n, observe that

$$f_1 = R_1 + R_2 + \cdots + R_n + f_{n+1}$$

and moreover, f_n tends to zero uniformly on $\mathbb{D}^-$. Thus the series

$$\sum_{n=1}^{\infty} R_n$$

is a Borel series equal to $f_1 = f$ on $\mathbb{D}$ and moreover, the sum of the absolute values of its coefficients does not exceed

$$\sum_{n=1}^{\infty} [R_n] \leq 2c,$$

$\square$

REMARK 4.2.8. As mentioned in the Overview, there are other results along these lines. For example, a result of Denjoy [42, 43] says that the coefficients $\{c_n\}$ in Wolff's theorem can be chosen to satisfy

$$|c_n| \leq k \exp(-n^{1/2-\varepsilon}).$$

Bonsall's example: Our next example of a non-trivial Borel series vanishing on one of its components is based on the following decomposition result of Bonsall which we state without proof.

PROPOSITION 4.2.9 (Bonsall [26]). *Let $\{z_n\} \subset \mathbb{D}$ be a sequence such that $|z_n| \to 1$ and almost every point of $\mathbb{T}$ is the nontangential limit of a subsequence of $\{z_n\}$. Then, given any $u \in L^1 = L^1(\mathbb{T}, d\theta)$, there is an absolutely summable sequence of complex numbers $\{c_n\}$ such that*

(4.2.10)
$$u = \sum_{n=1}^{\infty} c_n P_{z_n},$$

where

$$P_z(e^{it}) = \frac{1}{2\pi} \frac{1 - |z|^2}{|z - e^{it}|^2}$$

is the usual Poisson kernel, and the convergence in eq.(4.2.10) is in L^1.

REMARK 4.2.11. Such sequences $\{z_n\}$ satisfying the hypothesis of the proposition are called 'dominating sequences' for H^∞ and exist in abundance. For example [**32**, p. 172], given any sequence $r_n \to 1^-$, $c > 0$ (large), and an function $\phi : \mathbb{N} \to \mathbb{N}$ satisfying $\phi(n) > c/1 - r_n$, the desired sequence is formed by taking $\phi(n)$ equally spaced points on the circle $|z| = r_n$. Routine geometry shows that every point $e^{i\theta}$ on the circle can be approached inside an angle of opening 2β (placed symmetrically about the radius from the origin to $e^{i\theta}$) for β satisfying $\tan \beta > \pi/c$. If one is willing to do slightly more work, one can even make $\{z_n\}$ be the zero set of a non-trivial analytic function on $\mathbb{D}$ subject to certain growth restrictions, see Proposition 6.6.4. See also Theorem 4.2.12 (below) for another characterization of dominating sequences for H^∞ that better explains the use of the term 'dominating'.

Let us now, with the aid of Bonsall's result, construct a nontrivial Borel series summing to zero on $\mathbb{D}_e$. Let a be a point of $\mathbb{D}$ distinct from the (dominating) sequence $\{z_n\}$, and choose the u in eq.(4.2.10) to be P_a. Thus

$$P_a = \sum_{n=1}^{\infty} c_n P_{z_n}$$

for some $c_n = c_n(a)$ which are absolutely summable. The convergence here is in the $L^1(\mathbb{T})$ norm. If now h is any complex valued harmonic function in $\mathbb{D}$ with continuous extension to $\mathbb{D}^-$ we get, after multiplying both sides of the last equation by $h(e^{it})dt$ and integrating over $[0, 2\pi)$,

$$h(a) = \sum_{n=1}^{\infty} c_n h(z_n).$$

For example, $h(z) = 1/(z - w)$, where w is a point in $\mathbb{D}_e$, is a valid choice in the last equation and yields

$$\frac{1}{a - w} = \sum_{n=1}^{\infty} \frac{c_n}{z_n - w},$$

which after transposition of the left hand member to the right side, is the desired example.

The work of Brown, Shields, and Zeller: As the reader has probably already noticed, the ability to find sequences $\{z_n\} \subset \mathbb{D}$ clustering non-tangentially at almost every boundary point plays a crucial role in creating the Wolff-type counterexamples. A paper of Brown, Shields, and Zeller brings this idea to fruition.

THEOREM 4.2.12 (Brown, Shields, and Zeller [**32**]). *Let $\{z_n\}$ be a sequence of distinct points in $\mathbb{D}$ with $|z_n| \to 1$. The following are equivalent.*

(1) *There is an absolutely summable sequence $\{c_n\}$ of complex constants, not all zero, such that*

$$\sum_{n=1}^{\infty} \frac{c_n}{z - z_n} = 0 \ \text{for all} \ z \in \mathbb{D}_e.$$

(2) *For any bounded analytic function f on $\mathbb{D}$*

$$\sup\{|f(z)| : z \in \mathbb{D}\} = \sup\{|f(z_n)| : n \in \mathbb{N}\}.$$

(3) *For almost every θ, the point $e^{i\theta}$ may be approached non-tangentially by points of $\{z_n\}$.*

More about Borel series: For a sequence of points of $\{z_n\} \subset \mathbb{D}$ which satisfy the Blaschke condition[2]

$$\sum_{n=1}^{\infty}(1 - |z_n|) < \infty,$$

and an absolutely summable sequence $\{c_n\}$, form the Borel series

$$(4.2.13) \qquad f(z) = \sum_{n=1}^{\infty}\frac{c_n}{z - z_n}.$$

Under these hypothesis, we will show the following 'coherence' theorem for $f|\mathbb{D}$ and $f|\mathbb{D}_e$.

PROPOSITION 4.2.14. *For the Borel series f in eq.(4.2.13),*

(1) $f|\mathbb{D}$ *is a meromorphic function of bounded type.*
(2) $f|\mathbb{D}_e \in H^p(\mathbb{D}_e)$ *for each $0 < p < 1$, that is to say,*

$$(4.2.15) \qquad \sup_{r>1}\int_0^{2\pi}|f(re^{i\theta})|^p d\theta < \infty$$

for each $0 < p < 1$. Equivalently, $f(1/z)$, $z \in \mathbb{D}_e$, belongs to H^p.
(3) *For almost every $e^{i\theta}$,*

$$\lim_{r \to 1^-} f(re^{i\theta}) = \lim_{r \to 1^+} f(re^{i\theta}).$$

In the language of Definition 6.2.1, $f|\mathbb{D}$ and $f|\mathbb{D}_e$ are 'pseudocontinuations' of each other. Compare this example to Poincaré's example (Chapter 3). The idea of the proof was communicated to us by D. Khavinson. It follows a similar argument used in [**89**, cf. Sect. 4, 5].

PROPOSITION 4.2.16. *For the Borel series f in eq.(4.2.13) and*

$$r \in (0, \infty)\backslash\{|z_n|\},$$

$$\int_0^{2\pi}|f(re^{i\theta})|^p d\theta \le A_p\left\{\sum_{n=1}^{\infty}|c_n|\right\}^p, \quad \text{for all } 0 < p < 1.$$

PROOF. The estimate in the proposition for $r > 1$ follows from Theorem 3.2.6 since $f|\mathbb{D}_e$ is the Cauchy transform of the measure

$$d\mu = \sum_{n=1}^{\infty}c_n\delta_{z_n},$$

where δ_{z_n} is the unit point mass at z_n and $\|\mu\| = \sum|c_n| < \infty$.

To get the estimate for $0 < r < 1$, we proceed as follows: For $N \in \mathbb{N}$, let

$$f_N = \sum_{n=1}^{N}\frac{c_n}{z - z_n}.$$

---

[2]Such a sequence is definitely not a dominating sequence for H^∞ since a Blaschke sequence can be the zeros of a non-trivial H^∞ function (see condition (2) of Theorem 4.2.12).

For fixed $r \in E := (0,1) \backslash \{|z_n|\}$, we infer from the Hahn-Banach theorem that the distance from $f_N(re^{i\theta})$ to the space $H^1(\mathbb{T})$ (the boundary functions for H^1 functions) is equal to

$$\sup \left\{ \left| \int_0^{2\pi} f_N(re^{i\theta}) g(e^{i\theta}) e^{i\theta} \frac{d\theta}{2\pi} \right| : g \in H^\infty, |g| \leq 1 \right\}.$$

By the Cauchy integral formula, the quantity inside the supremum is equal to

$$\frac{1}{r} \left| \sum_{n \leq N, |z_n|/r < 1} c_n \, g(z_n/r) \right|$$

which is bounded above by $2 \sum_{n \geq 1} |c_n|$, assuming, without loss of generality, that $r > 1/2$.

Thus, for each $N \in \mathbb{N}$, there is an $h_N \in H^1$ with

$$\int_0^{2\pi} |f_N(re^{i\theta}) - h_N(e^{i\theta})| \frac{d\theta}{2\pi} \leq C \sum_{n=1}^\infty |c_n|,$$

where C is independent of N. By Smirnov's theorem (Theorem 3.2.6) the H^p norm of the following Cauchy transform

$$(4.2.17) \qquad \int_0^{2\pi} \frac{f_N(re^{i\theta}) - h_N(e^{i\theta})}{e^{i\theta} - z} e^{i\theta} \frac{d\theta}{2\pi}$$

is bounded above by

$$A_p \int_0^{2\pi} |f_N(re^{i\theta}) - h_N(e^{i\theta})| d\theta \leq A_p \sum_{n=1}^\infty |c_n|.$$

By Cauchy's theorem,

$$\int_0^{2\pi} \frac{f_N(re^{i\theta})}{e^{i\theta} - z} e^{i\theta} \frac{d\theta}{2\pi} = \frac{1}{r} \sum_{n \leq N, |z_n| > r} \frac{c_n}{z - z_n/r} \quad \text{for all } |z| < 1$$

and an application of the Cauchy integral formula shows that

$$\int_0^{2\pi} \frac{h_N(e^{i\theta})}{e^{i\theta} - z} e^{i\theta} \frac{d\theta}{2\pi} = h_N(z) \quad \text{for all } |z| < 1.$$

Combining these two results gives us

$$h_N(z) = \int_0^{2\pi} \frac{h_N(e^{i\theta}) - f_N(re^{i\theta})}{e^{i\theta} - z} e^{i\theta} \frac{d\theta}{2\pi} + \frac{1}{r} \sum_{n \leq N, |z_n| > r} \frac{c_n}{z - z_n/r}.$$

From eq.(4.2.17) (applied to the first term) and Smirnov's theorem (applied to the second term), the H^p norm of h_N does not exceed

$$A_p \sum_{n=1}^\infty |c_n|.$$

Notice that

$$\int_0^{2\pi} |f_N(re^{i\theta})|^p d\theta \leq \int_0^{2\pi} |f_N(re^{i\theta}) - h_N(e^{i\theta})|^p d\theta + \int_0^{2\pi} |h_N|^p d\theta$$

$$\leq A_p \left\{ \int_0^{2\pi} |f_N(re^{i\theta}) - h_N(e^{i\theta})| d\theta \right\}^p + A_p \left\{ \sum_{n \geq 1} |c_n| \right\}^p$$

$$\leq A_p \left\{ \sum_{n \geq 1} |c_n| \right\}^p .$$

The conclusion of the proposition now follows (for $r < 1$) since

$$f_N(re^{i\theta}) \to f(re^{i\theta})$$

uniformly in θ as $N \to \infty$ (note that $r \in E = (0,\infty)\setminus\{|z_n| : n \in \mathbb{N}\}$). □

The first two parts of Proposition 4.2.14 are contained in the following Corollary to Proposition 4.2.16.

COROLLARY 4.2.18. *For the above Borel series f in eq.(4.2.13),*

(1) $f|\mathbb{D}$ *is a meromorphic function of bounded type,*
(2) $f|\mathbb{D}_e \in H^p(\mathbb{D}_e)$ *for each $0 < p < 1$.*

PROOF. Statement (2) follows from Proposition 4.2.16. To prove statement (1), let b be a Blaschke product with zeros $\{z_n\}$ and note that bf is an analytic function on $\mathbb{D}$ which is the uniform limit (on compact subsets of $\mathbb{D}$) of the functions bf_N as $N \to \infty$. By Proposition 4.2.16, the H^p norms of bf_N are uniformly bounded for each fixed $0 < p < 1$ and so $bf \in H^p$. This means that $f|\mathbb{D}$ is the quotient of an H^p function and a bounded analytic function from which follows that $f|\mathbb{D}$ is a function of bounded type. □

Since $f|\mathbb{D}$ and $f|\mathbb{D}_e$ are functions of bounded type (on their respective domains) we know that the radial (and even non-tangential) limits exist almost everywhere. This next result shows that they are indeed equal almost everywhere, proving part (3) of Proposition 4.2.14.

PROPOSITION 4.2.19. *For the Borel series f as in eq.(4.2.13),*

$$\lim_{r \to 1^-} f(re^{i\theta}) = \lim_{r \to 1^+} f(re^{i\theta}) \quad a.e.$$

PROOF. We first claim that for each fixed $0 < p < 1$.

$$(4.2.20) \qquad \int_0^{2\pi} |f(re^{i\theta}) - f(e^{i\theta}/r)|^p d\theta \to 0 \quad \text{as } r \to 1^-, \ r \neq |z_n| \ (n = 1,2,\ldots).$$

To see this, let $\varepsilon > 0$ be given and choose N large enough so that $\sum_{n>N} |c_n| < \varepsilon$. Write

$$f = \sum_{n=1}^N \frac{c_n}{z - z_n} + \sum_{n=N}^\infty \frac{c_n}{z - z_n} = f_N + g_N.$$

The integral in eq.(4.2.20) is bounded above by

$$\int_0^{2\pi} |f_N(re^{i\theta}) - f_N(e^{i\theta}/r)|^p d\theta + \int_0^{2\pi} |g_N(re^{i\theta}) - g_N(e^{i\theta}/r)|^p d\theta.$$

The first integral goes to zero as $r \to 1^-$ since f_N is continuous near the circle. By Proposition 4.2.16, the second integral is bounded above by

$$A_p \left\{ \sum_{n>N} |c_n| \right\}^p < A_p \varepsilon^p.$$

This verifies eq.(4.2.20).

Since $f|\mathbb{D}_e \in H^p(\mathbb{D}_e)$, then as $r \to 1^-$, $f(e^{i\theta}/r)$ approaches its (exterior) boundary function, call it $h(e^{i\theta})$ both in L^p norm as well as pointwise almost everywhere. By eq.(4.2.20), $f(re^{i\theta}) \to h(e^{i\theta})$ in L^p norm and hence in measure. But since $f|\mathbb{D}$ is of bounded type, then it has finite radial limits almost everywhere, and hence in measure. This means that $f(re^{i\theta}) \to h(e^{i\theta})$ almost everywhere and the result follows. □

The work of Beurling: We have already seen that the two component functions $f|\mathbb{D}$ and $f|\mathbb{D}_e$ of the superconvergent Borel series

$$f(z) = \sum_{n=1}^\infty \frac{c_n}{z - z_n}, \quad |z_n| \to 1, \quad \sqrt[n]{|c_n|} \to 0,$$

enjoy certain coherence properties. As mentioned earlier, there are results of Beurling, which we will now discuss, and those of Gončar, which will be discussed in the next chapter, that obtain the same coherence results but with a weaker hypothesis than $\sqrt[n]{|c_n|} \to 0$.

The ideas of Beurling [**18, 22**] are the following: Consider a Borel series of the form

$$(4.2.21) \qquad \sum_{n=1}^\infty \frac{c_n}{z - z_n},$$

where $\{c_n\}$ is an absolutely convergent sequence of complex numbers but $\{z_n\}$ can be *any* sequence of points in the plane. They can, for example, be a dense set in the plane, giving the reader pause as to whether or not the above Borel series converges at all. This is resolved with the following proposition.

PROPOSITION 4.2.22. *The Borel series in eq.(4.2.21) converges absolutely for almost every (with respect to area measure in the plane) $z \in \mathbb{C}$.*

PROOF. For each $R > 0$, it is easy to check that for any $a \in \mathbb{C}$

$$\int_{\{|z|<R\}} \frac{dA(z)}{|z - a|} \le 2\pi R$$

(dA denotes area measure in the plane) and so

$$\int_{\{|z|<R\}} \sum_{n=1}^\infty \frac{|c_n|}{|z - z_n|} dA(z) = \sum_{n=1}^\infty |c_n| \int_{\{|z|<R\}} \frac{dA(z)}{|z - z_n|} \le 2\pi R \sum_{n=1}^\infty |c_n|.$$

The result now follows. □

It is important to remind the reader that the Borel series Beurling discusses here need not be analytic anywhere in the plane. Nevertheless, we do have the following coherence result.

THEOREM 4.2.23 (Beurling). *If*

$$r_n := \sum_{j=n+1}^{\infty} |c_j|$$

satisfies

$$\varliminf_{n \to \infty} \sqrt[n]{r_n} < 1$$

and if the Borel series

$$\sum_{n=1}^{\infty} \frac{c_n}{z - z_n}$$

vanishes on a set of Hausdorff one-dimensional measure, then $c_n = 0$ for every n.

COROLLARY 4.2.24. *If*

$$|z_n| \downarrow 1 \quad and \quad \varlimsup_{n \to \infty} \sqrt[n]{|c_n|} < 1$$

and

$$f(z) := \sum_{n=1}^{\infty} \frac{c_n}{z - z_n}$$

vanishes on $\mathbb{D}$, then $c_n = 0$ for all n.

The above corollary demonstrates the coherence of $f|\mathbb{D}$ and $f|\mathbb{D}_e$ but not yet in the strongest possible sense. Later, work of Gončar (see the next chapter) proves a stronger result, namely (under the same hypothesis), if $f|\mathbb{D}$ has an analytic continuation across some point of $\mathbb{T}$, then this analytic continuation must be none other than $f|\mathbb{D}_e$.

Before moving on, we mention a more general rational approximation theorem of Beurling along the line of Theorem 4.2.23. Let $\{s_n\}$ be a sequence of positive numbers with $s_n \to 0$ as $n \to \infty$. For a connected open set $G \subset \mathbb{C}$, let $(G, \{s_n\})$ denote the class of almost everywhere (with respect to area measure in the plane) defined functions f which are almost everywhere limits of rational functions of the form $f_n = P_n/Q_n$ (where P_n and Q_n are analytic polynomials of degree at most n with no common roots) and for which

$$\text{Area}(\{z : |f(z) - f_n(z)| > s_n\}) \to 0 \quad \text{as } n \to \infty.$$

THEOREM 4.2.25 (Beurling). *The condition*

$$\varliminf_{n \to \infty} \sqrt[n]{s_n} < 1$$

is both necessary and sufficient for the class $(G, \{s_n\})$ to have the property that any two functions in this class which coincide on a set of positive area measure are in fact equal almost everywhere on G.

CHAPTER 5

Gončar Continuation

5.1. Hyperconvergence

Recall our set-up at the beginning of Chapter 4 on 'superconvergence': Let A be a compact set in the plane, and $\{P_n\}, \{Q_n\}$ be two sequences of (analytic) polynomials with complex coefficients satisfying

$$\deg P_n \leq n, \ \deg Q_n \leq n,$$

$$Q_n^{-1}(\{0\}) \subset A,$$

P_n and Q_n have no common roots.

Define

(5.1.1) $$R_n = P_n/Q_n.$$

Thus R_n is a rational function with at most n zeros in the plane and at most n poles in A. Recall (Definition 4.1.2) that a sequence $\{R_n\}$ is 'superconvergent' on a compact subset $K \subset A^c$ if

(5.1.2) $$\max \left\{ |R_n(z) - R_{n+1}(z)| : z \in K \right\}^{1/n} \to 0 \text{ as } n \to \infty.$$

One of the highlights of Chapter 4 was the theorem of Walsh (Proposition 4.1.11) which said that if $\{R_n\}$ is superconvergent on a non-degenerate continuum K, then the superconvergence propagates to every compact subset of A^c. The other highlight (Theorem 4.1.14) was the fact that if a superconvergent Borel series, for example,

$$\sum_{n=1}^{\infty} \frac{c_n}{z - z_n}, \quad \lim_{n \to \infty} \sqrt[n]{|c_n|} = 0,$$

has an analytic continuation, along any path, to any other disk, that analytic continuation must coincide with the sum of the Borel series. In other words, regarding the component functions $f|_{\Omega_k}$ and $f|_{\Omega_j}$, where $\{\Omega_j : j = 1, 2, \ldots\}$ are the connected components of $\mathbb{C} \backslash \{z_n\}^-$, as 'continuations' of each other is compatible with analytic continuation.

In this section, we treat the work of Gončar where the superconvergence condition in eq.(5.1.2) is relaxed to

$$\max \left\{ |R_n(z) - R_{n+1}(z)| : z \in K \right\} \leq C_K \rho^n, \ \rho < 1.$$

Although in a weaker form, the propagation result still holds. The main result of Gončar (see Theorem 5.2.3 below) says that compatibility with analytic continuation also holds (at least for adjacent components Ω_j and Ω_k). In particular, the same compatibility result holds for Borel series whose coefficients satisfy the weaker condition

$$\varlimsup_{n \to \infty} \sqrt[n]{|c_n|} < 1.$$

37

DEFINITION 5.1.3. For a sequence of rational functions $\{R_n\}$, as in eq.(5.1.1), a compact subset $K \subset A^c$, and $\rho < 1$, we say $\{R_n\}$ is 'ρ-hyperconvergent on K' if

$$\max \big\{ \, |R_{n+1}(z) - R_n(z)| : z \in K \, \big\} \leq C_K \rho^n \quad \text{for all } n = 1, 2, \ldots$$

Nearly exactly as before for 'superconvergent' rational functions (Proposition 4.1.4), one can prove the following.

PROPOSITION 5.1.4. *The following are equivalent.*

(1) $\{R_n\}$ *is ρ-hyperconvergent on K.*
(2) *There is an $f \in C(K) \cap \mathfrak{H}(K^o)$ such that*

$$\max \big\{ \, |f(z) - R_n(z)| : z \in K \, \big\} \leq C_K \rho^n.$$

From the Bernstein-Faber theroem (Theorem 4.1.9 and the remark following), we know that if $\{P_n\}$ is a sequence of polynomials, each of degree n, which ρ-hyperconverges on a non-degenerate continuum K, then for each number r satisfying $0 < \rho < r < 1$, there is an open set U_r containing K such that $\{P_n\}$ r-hyperconverges on U_r^-. We now prove a version of this propagation result for a ρ-hyperconvergent sequence of *rational* functions.

PROPOSITION 5.1.5. *Let $K \subset A^c$ be a non-degenerate continuum and $\{R_n\}$, as in eq.(5.1.1), be a ρ-hyperconvergent sequence of rational functions on K. Then for each number r satisfying $0 < \rho < r < 1$ there is an open neighborhood U_r containing K such that $\{R_n\}$ is r-hyperconvergent on U_r^-.*

For a non-degenerate continuum K and $D > 0$, let

$$G_D := \{z : \operatorname{dist}(z, K) < D\}.$$

LEMMA 5.1.6. *Suppose K and G_D are as above and f is a polynomial of degree at most N, all of whose zeros lie outside G_D. Then for $z \in G_D$,*

$$|f(z)| \geq m \exp \Big(-N \frac{d(z)}{D} \Big),$$

where $m = \min\{|f(z)| : z \in K\}$ and $d(z) = dist(z, K)$.

PROOF. Let $n \leq N$ be the degree of f and $z_1, \ldots, z_n$ denote the zeros of f. For $z \in G_D \backslash K$ let w be one of the nearest points in K to z, i.e.,

(5.1.7) $|w - z| = d(z)$.

One easily argues from the definition of G_D, that G_D is connected and moreover, the line segment from z to w lies wholly in G_D. The function $\log f(z)$ may be multiple valued in G_D, since G_D may not be simply connected. However, it is locally analytic and has a single-valued derivative f'/f and a single-valued real part $\log |f|$.

Note that for $z \in G_D$,

$$\frac{f'(z)}{f(z)} = \sum_{j=1}^{n} \frac{1}{z - z_j}$$

and so

(5.1.8)
$$\left| \frac{f'(z)}{f(z)} \right| \leq \frac{n}{D}.$$

For some branch of the logarithm,

$$\log f(w) - \log f(z) = \int_z^w \frac{f'(t)}{f(t)} dt.$$

Taking real parts and using eq.(5.1.8) and eq.(5.1.7) we get

$$\log |f(w)| - \log |f(z)| = \Re \left\{ \int_z^w \frac{f'(t)}{f(t)} dt \right\} \leq \frac{n}{D} d(z).$$

Thus

$$|f(w)| \leq |f(z)| \exp\left(\frac{n}{D} d(z) \right).$$

Hence, by definition of m, being the minimum of $|f|$ on K, and the fact that $n \leq N$, we get

$$|f(z)| \geq m \exp\left(-\frac{n}{D} d(z) \right) \geq m \exp\left(-\frac{N}{D} d(z) \right).$$

$\square$

LEMMA 5.1.9. *Let K be a non-degenerate continuum contained in a closed disk D_1 of radius R_1 and f be a polynomial of degree at most N having no zeros interior to a larger concentric disk D_2 of radius R_2. Finally, let*

$$M = \max\{|f(z)| : z \in K\} \quad and \quad m = \min\{|f(z)| : z \in K\}.$$

Then

$$\frac{M}{m} \leq \left(\frac{R_2 + R_1}{R_2 - R_1} \right)^N.$$

PROOF. Let $n \leq N$ be the degree of f. Since we are estimating M/m, we can assume, without loss of generality, that f has as leading coefficient equal to one. Thus

$$f(z) = (z - z_1) \cdots (z - z_n),$$

where $z_1, \ldots, z_n$ are the zeros of f, which by assumption satisfy $|z_j| > R_2$. We also assume the disks D_1 and D_2 are centered at $z = 0$.

Let a be a point of K where $|f|$ attains its maximum value M, and b a point of K where $|f|$ attains its minimum m. Then,

$$M \leq (|a| + |z_1|) \cdots (|a| + |z_n|) \quad and \quad m \geq (|z_1| - |b|) \cdots (|z_n| - |b|).$$

Hence, M/m does not exceed

$$\frac{|z_1| + |a|}{|z_1| - |b|} \cdots \frac{|z_n| + |a|}{|z_n| - |b|}.$$

The function $t \to (t + |a|)/(t - |b|)$ is monotone decreasing on (R_2, ∞) and so it is bounded above there by

$$\frac{R_2 + |a|}{R_2 - |b|} \leq \frac{R_2 + R_1}{R_2 - R_1}.$$

which, after remembering that $n \leq N$ and $(R_2 + R_1)/(R_2 - R_1) \geq 1$, implies the result.

$\square$

LEMMA 5.1.10. *For each non-degenerate continuum K, $w \notin K$, and polynomial f of degree at most N,*

$$(5.1.11) \qquad |f(w)| \leq A(K,w)^N \sup\{|f(z)| : z \in K\},$$

where $A(K,w) > 1$ is a constant that, for fixed K, is continuous in w and tends to 1 as $w \to K$.

PROOF. If $\hat{K}$ is the polynomial convex hull of K, that is to say, the set of $z \in \mathbb{C}$, such that

$$|p(z)| \leq \sup\{|p(w)| : w \in K\}$$

for all polynomials p, then $\hat{K}$ is a non-degenerate continuum and moreover $(\hat{K})^c$ is the unbounded component of K^c. Thus to prove eq.(5.1.11), we can assume that $\hat{K} = K$ (*i.e.*, K is 'polynomially convex'). Since K (which we now assume is equal to $\hat{K}$) is a non-degenerate continuum, then K^c admits a Green's function $g(z)$ with pole at infinity which grows like

$$\log|z| + c + O(|z|^{-1}), \quad |z| \to \infty.$$

Furthermore, $g(z) \to 0$ as $z \to \partial K$. See [**139**, Thm. I.11, Thm. I.18, Thm. III.35] for further details.

If $f = az^n + \cdots$ is a polynomial of degree $n \leq N$ with $|f| \leq M$ on K, then $\log|f(z)|$ is subharmonic on $\mathbb{C}$ with growth at infinity equal to

$$n \log|z| + \log|a| + O(|z|^{-1}).$$

Thus the function

$$u(z) := \log|f(z)| - ng(z)$$

is subharmonic on K^c and bounded near infinity and so, by the maximum principle,

$$\log|f(z)| - ng(z) \leq \log M \quad \text{for all } z \in K^c.$$

From this follows

$$|f(z)| \leq M(e^{g(z)})^n \leq M(e^{g(z)})^N$$

which, letting $A(K,z) = e^{g(z)}$, is the desired result. $\qquad\square$

PROOF OF PROPOSITION 5.1.5. Since $K \subset A^c$, there is a $D > 0$ such that

$$G_D = \{z : \text{dist}(z,K) < D\} \subset A^c.$$

For each $n \in \mathbb{N}$, the rational function

$$R_n - R_{n+1} = \frac{f_n}{g_n},$$

where f_n and g_n are polynomials of degrees at most $2n+1$ and g_n has no zeros in G_D.

If

$$M_n = \max\{|g_n(z)| : z \in K\} \quad \text{and} \quad m_n = \min\{|g_n(z)| : z \in K\},$$

then using the ρ-hyperconvergence of R_n, we obtain the estimate

$$|f_n(z)| \leq C_K M_n \rho^n \quad \text{for all } z \in K.$$

By Lemma 5.1.10,

$$|f_n(z)| \leq A(K, K_d)^{2n+1} C_K M_n \rho^n \quad \text{for all } z \in K_d,$$

where, for $0 < d < D$,
$$K_d = \{z : \operatorname{dist}(z, K) \le d\}.$$
Here, $A(K, K_d) > 1$ and approaches one as $d \to 0$. From Lemma 5.1.6,
$$|g_n(z)| \ge m_n e^{-(2n+1)d/D}, \quad z \in K_d,$$
and so
$$(5.1.12) \qquad |R_n(z) - R_{n+1}(z)| \le C_K \frac{M_n}{m_n} A(K, K_d)^{2n+1} e^{(2n+1)d/D} \rho^n, \quad z \in K_d.$$
Repeat the above argument with the continuum K replaced by the smaller continuum
$$K(s, w) = K \cap \{|z - w| \le s\},$$
where $w \in K$ and $0 < s < D$. For this new continuum, apply Lemma 5.1.9 with $D_1 = \{|z - w| \le s\}$ and $D_2 = \{|z - w| \le D\}$ to obtain the estimate
$$\frac{M_n}{m_n} \le \left(\frac{1 + s/D}{1 - s/D} \right)^{2n+1},$$
where M_n and m_n are understood to be the max and min of $|g_n|$ on $K(s, w)$. First choose a small enough s so that
$$\left(\frac{1 + s/D}{1 - s/D} \right)^{2n+1} \rho^n \le r_1^n, \quad n = 1, 2, \ldots$$
for some r_1 with $\rho < r_1 < 1$. For this fixed s, apply eq.(5.1.12), with K replaced by $K(s, w)$, and choose a small enough d so that
$$A(K, K_d)^{2n+1} e^{(2n+1)d/D} r_1^n \le r_2^n, \quad n = 1, 2, \ldots$$
for some r_2 with $r_1 < r_2 < 1$. At least on the set $K(s, w)_d$, $\{R_n\}$ r_2-hyperconverges. The result now follows from a compactness argument. □

5.2. Gončar continuation

DEFINITION 5.2.1. Let E be a non-empty closed subset of $\mathbb{T}$. We say that $f_1 \in \mathfrak{M}(\Omega_1)$, where Ω_1 is a region (a connected open set) in $\mathbb{D}$ with $\Omega_1^- \supset E$, admits a 'Gončar continuation' (abbreviated GC) across E if there is

(1) a region $\Omega_2 \subset \mathbb{D}_e$ with $\Omega_1^- \cap \Omega_2^- \supset E$.
(2) an $f_2 \in \mathfrak{M}(\Omega_2)$
(3) a compact set A, containing the poles of f_1 and f_2, with no limit points in $\Omega_1 \cup \Omega_2$.
(4) a number $\rho \in (0, 1)$
(5) a sequence of rational functions $\{R_n\}$, as in eq.(5.1.1), such that
    (a) $\{R_n\}$ ρ-hyperconverges to f_1 on compact subsets of $\Omega_1 \backslash A$.
    (b) $\{R_n\}$ ρ-hyperconverges to f_2 on compact subsets of $\Omega_2 \backslash A$.

EXAMPLE 5.2.2. Certainly any Borel series
$$F(z) = \sum_{n=1}^{\infty} \frac{A_n}{z - z_n},$$
where the sequence $\{z_n\}$ has no accumulation points in $\mathbb{C}_\infty \backslash \mathbb{T}$ and
$$\varlimsup_{n \to \infty} \sqrt[n]{|A_n|} < 1,$$

has a Gončar continuation (GC) across $\mathbb{T}$.

We have seen earlier that a Borel-Walsh continuation is compatible with analytic continuation (see the remarks just before Definition 4.2.1), that is to say, if $f \in \mathfrak{H}(A^c)$ is the limit of a superconvergent sequence of rational functions with poles in A, and $f|\Omega_j$ has an analytic continuation along some path to a point in Ω_k (here Ω_j and Ω_k are two components of A^c), then this analytic continuation must be equal to $f|\Omega_k$. This next result of Gončar [**62, 63**] shows the same is true for GC, at least when Ω_j and Ω_k are adjacent components. To make the exposition clear and to avoid needless geometric technicalities, we state and prove the following version of Gončar's theorem.

THEOREM 5.2.3 (Gončar). *Suppose A is a compact subset of $\mathbb{D}_e^-$ with no accumulation points in $\mathbb{D}_e$ and that $\{R_n\}$, as in eq.(5.1.1), is a sequence of rational functions which ρ-hyperconverge to $f \in \mathfrak{H}(\mathbb{D})$ on compact subsets of $\mathbb{D}$. Furthermore, assume that R_n converges uniformly on compact subsets of $\mathbb{D}_e \backslash A$ to a $g \in \mathfrak{M}(\mathbb{D}_e)$. If f has an analytic continuation, also denoted by f, to a neighborhood U of $e^{i\theta}$, then $f = g$ on $U \cap \mathbb{D}_e$.*

We shall deduce this from the following result of Gončar.

PROPOSITION 5.2.4. *Suppose, in Theorem 5.2.3, that f is analytically continuable to*
$$S := \{z : |\arg z| < \alpha, |z| < R\}$$
for some $\alpha > 0$ and $R > 1$. Then, there is a number $b > 1$ such that each arc
$$A(r) := \{z : |z| = r, |\arg z| \le \alpha/2\}$$
with fixed $1 < r < b$ satisfies
$$\min\{|f(z) - R_n(z)| : z \in A(r)\} \to 0 \ \ as \ n \to \infty.$$

Assuming this proposition for a moment, Theorem 5.2.3 can be shown as follows: We have, by virtue of Proposition 5.2.4, a sequence $\{w_n\}$ on the arc $A(r)$ such that $f(w_n) - R_n(w_n) \to 0$ as $n \to \infty$. Passing to a subsequence if necessary, we may assume w_n tend to a point w of $A(r)$. Since our assumptions imply that $R_n \to g$ uniformly on some neighborhood of w, we have $f(w) - g(w) = 0$, *i.e.*, f and g are equal at a point of $A(r)$. Since this is true for all r in some interval, they are equal on an uncountable subset of $S \backslash A$, and hence they are identical.

Before proceeding to the proof of Proposition 5.2.4, we need the following fundamental estimate.

LEMMA 5.2.5. *Let $R = P/Q$ where P and Q are polynomials of degree at most n without common zeros. Let K and L be disjoint non-degenerate continua in $\mathbb{C}$. Then, there is a constant $M(K, L)$ such that*

(5.2.6) $\min\{|R(z)| : z \in L\} \le M(K,L)^n \max\{|R(z)| : z \in K\}.$

Moreover, if $diam(K)$ and $diam(L)$ each exceed d, and L is included in an ε-neighborhood of K, and K is included in an ε-neighborhood of L, then $M(K,L)$ is bounded by a number that tends to 1 as $\varepsilon \to 0$ (for d fixed), that is to say, $\sup M(K,L)$ over all pairs K, L satisfying $diam(K) \ge d$, $diam(L) \ge d$ and such *that each of K and L is contained in an ε-neighborhood of the other, tends (for fixed d) to 1 as $\varepsilon \to 0$.*

PROOF. There is no loss of generality to assume (because of the homogeneity in eq.(5.2.6)),

$$\text{(5.2.7)} \qquad \max\{|R(z)| : z \in K\} = 1.$$

Also, writing tP and tQ in place of P and Q we may assume, by suitably choosing t, that

$$\text{(5.2.8)} \qquad \max\{|P(z)| : z \in L\} = 1.$$

Let $m := \min\{|R(z)| : z \in L\}$. Thus,

$$|P(z)| \geq m|Q(z)| \text{ for all } z \in L.$$

Hence, in view of eq.(5.2.8) and Lemma 5.1.10,

$$|Q(z)| \leq \frac{A(L,z)^n}{m} \text{ for all } z \in \mathbb{C}.$$

In particular,

$$|Q(z)| \leq \frac{B(L,K)^n}{m} \text{ for all } z \in K,$$

where $B(L,K)$ denotes $\max\{A(L,z) : z \in K\}$.

Hence, using eq.(5.2.7), we see that for $z \in K$,

$$|P(z)| \leq \frac{B(L,K)^n}{m}.$$

Again, from Lemma 5.1.10, for $z \in L$ we have

$$\text{(5.2.9)} \qquad |P(z)| \leq \frac{1}{m}A(K,z)^n B(L,K)^n \leq \frac{1}{m}B(K,L)^n B(L,K)^n$$

where $B(K,L)$ denotes $\max\{A(K,z) : z \in L\}$.

Finally, in view of eq.(5.2.8), $|P(a)| = 1$ for some $a \in L$. Taking $z = a$ in eq.(5.2.9) yields $m \leq M(K,L)^n$, where $M(K,L) := B(K,L)B(L,K)$. This proves eq.(5.2.6). The remaining assertions follow from knowledge about $B(K,L)$ as a function of K and L, which are implicit in eq.(5.1.11). □

PROOF OF PROPOSITION 5.2.4. Choose a number s, less than but close to one and let

$$B(s) := \{z : |z| = s, |\arg z| \leq \alpha/2\},$$

and let $B(1/s)$ denote the corresponding concentric arc on $\{|z| = 1/s\}$.

Since f is analytically continuable to $S = \{z : |\arg z| < \alpha, |z| < R\}$, by the Bernstein-Faber theorem (Theorem 4.1.9), there is a sequence of polynomials $\{P_n\}$, each of degree n, such that

$$\max\{|f(z) - P_n(z)| : |\arg z| \leq \alpha/2, |z| \leq R - \varepsilon\} \leq Ct^n$$

for some constant $t < 1$ and constant C, and this t is independent of s. Also, by hypothesis (the ρ-hyperconvergence of R_n to f on compact subsets of $\mathbb{D}$), we have

$$\max\{|f(z) - R_n(z)| : z \in B(s)\} \leq C(s)\rho^n,$$

where $C(s)$ is some constant, which may depend on s. From the last two estimates,

$$|R_n(z) - P_n(z)| \leq C(s)(t^n + \rho^n),$$

for $z \in B(s)$, with some constant $C(s)$ that may depend on s (but not n).

Now, $R_n - P_n$ is a rational function with at most $2n$ zeros and at most n poles. By the 'fundamental estimate' in eq.(5.2.6), the minimum of its modulus on $B(1/s)$ does not exceed the maximum of its modulus on $B(s)$, times a factor M^{2n}, where M is a positive constant depending only on the geometric configuration of the two continua $B(s)$ and $B(1/s)$. In view of remarks made earlier, this M can be chosen arbitrarily close to 1 when s is sufficiently near 1. Thus we have

$$\min\{|R_n(z) - P_n(z)| : z \in B(1/s)\} \leq M^{2n} C(s)(t^n + \rho^n).$$

If s is close enough to 1, we will have M^2 less than both $1/t$ and $1/\rho$ and consequently

$$\min\{|R_n(z) - P_n(z)| : z \in B(1/s)\} \to 0 \quad \text{as } n \to \infty.$$

Since, moreover, $P_n(z) - f(z)$ tends to zero uniformly for z in $B(1/s)$ as $n \to \infty$, the conclusion of the proposition follows.   □

CHAPTER 6

Pseudocontinuation

6.1. A problem of Walsh and Tumarkin

In 1929, J. L. Walsh considered the following rational approximation problem: For each $n \in \mathbb{N}$, let

$$S_n := \{z_{n,1}, z_{n,2}, \ldots, z_{n,N(n)}\}$$

denote a finite sequence of points in the extended exterior disk $\mathbb{D}_e$. The point at ∞ is allowed to be one of the points and the points may be repeated. Let

$$R_n := \bigvee \left\{ \frac{1}{z - z_{n,j}}, \ j = 1, \ldots, N(n) \right\},$$

where $\bigvee$ denotes the 'closed linear span'. If, for example, the point $z_{n,j} \neq \infty$ appears k times in the sequence S_n, then R_n also includes the functions

$$\frac{1}{(z - z_{n,j})^s}, \quad s = 1, \ldots, k$$

in its spanning set. If ∞ appears k times in the sequence S_n, the functions $1, z, \ldots, z^{k-1}$ are also included in the spanning set for R_n.

QUESTION 6.1.1. Under what conditions on the tableau $\{S_n : n \in \mathbb{N}\}$ is it true that *every* function f in the Hardy space H^2 (recall from Chapter 2 the definition of H^2 and its basic properties) can be represented as

$$f = \lim_{n \to \infty} f_n,$$

where $f_n \in R_n$ for each n and the limit is in the H^2-norm, that is to say,

$$\|f - f_n\|_{H^2}^2 = \int_0^{2\pi} |f(e^{i\theta}) - f_n(e^{i\theta})|^2 \frac{d\theta}{2\pi} \to 0 \quad \text{as } n \to \infty?$$

Actually, Walsh considered more general problems with the poles $z_{n,j}$ chosen from both $\mathbb{D}$ and $\mathbb{D}_e$, and the functions to be approximated were in $L^2(\mathbb{T}, d\theta)$ rather than H^2, but, for our purposes, the essential features of his investigation become sufficiently clear from the examination of the above special case. He proved the following.

THEOREM 6.1.2 (Walsh [143]). *For a tableau $\{S_n : n \in \mathbb{N}\}$ as above, the following are equivalent:*

(1) *Every $f \in H^2$ can be written as*

$$f = \lim_{n \to \infty} f_n, \quad f_n \in R_n.$$

(2) *The numbers*

$$p_n := \sum_{j=1}^{N(n)} (1 - |z_{n,j}|^{-1})$$

tend to infinity as $n \to \infty$.

We shall shortly provide a proof of this. Walsh did not, however, characterize the subset $\mathcal{M}$ of H^2 consisting of (norm) limits of sequences $\{f_n : f_n \in R_n\}$ in the case when

$$\varliminf_{n \to \infty} p_n < \infty.$$

This question is close to our central theme, because it is easily seen that $\mathcal{M}$ is a closed subspace of H^2 invariant under the backward shift operator

$$f \to \frac{f - f(0)}{z}.$$

Indeed, R_n consists of linear combinations of eigenvectors and root vectors of the backward shift (see Chapter 6, § 6.5). Among other things, Tumarkin proved in a series of fundamental papers, which foreshadow more recent results concerning pseudocontinuations and cyclic vectors for the backward shift, the following result.

THEOREM 6.1.3 (Tumarkin [**140**]). *Suppose, with the above notations,*

$$\varliminf_{n \to \infty} p_n < \infty.$$

Then, for every $f \in H^2$ *which can be written as*

$$f = \lim_{n \to \infty} f_n, \quad f_n \in R_n,$$

there are bounded analytic functions F *and* G *on* $\mathbb{D}_e$ *such that*

(6.1.4) $$\lim_{r \to 1^-} f(re^{i\theta}) = \lim_{r \to 1^+} \frac{F}{G}(re^{i\theta}) \quad a.e. \ \theta.$$

Moreover, every such $f \in H^2$, *for which there are bounded analytic* F *and* G *on* $\mathbb{D}_e$ *satisfying eq.(6.1.4), can be obtained as*

$$f = \lim f_n, \quad f_n \in R_n,$$

for some choice of tableau $\{S_n : n \in \mathbb{N}\}$ *with* $\{p_n\}$ *bounded.*

REMARK 6.1.5. A proof of Tumarkin's theorem can be found in [**103**, p. 37].

Later, in a paper of Douglas, Shapiro, and Shields [**51**], it became clear how this result is clarified by the description of the non-cyclic vectors for the backward shift. We will certainly say more about this (Theorem 6.3.4), but first we prove the Walsh result. We will only prove the sufficiency part. The necessity follows in a sharper form from Tumarkin's result.

PROOF OF WALSH'S THEOREM. Suppose $p_n \to \infty$. It will be convenient to write

$$a_n = 1/\overline{z_n}$$

and rephrase the problem at hand in terms of the reproducing kernels

$$k_w(z) = \frac{1}{1 - \overline{w}z}, \quad w \in \mathbb{D}.$$

of H^2. Note that by the Cauchy integral formula, $\langle f, k_w \rangle = f(w)$ for $w \in \mathbb{D}$. The rephrased problem is the following: For $n \in \mathbb{N}$, let

$$T_n := \{a_{n,1}, a_{n,2}, \ldots, a_{n,N(n)}\} \subset \mathbb{D} \text{ (repetitions allowed)}$$

be chosen such that

$$p_n := \sum_{j=1}^{N(n)} (1 - |a_{n,j}|) \to \infty \text{ as } n \to \infty.$$

Let

$$K_n := \bigvee \{ k_w : w \in T_n \}.$$

If some number a appears multiply in T_n, say r times, then include the functions

$$k_a, k_a', \ldots, k_a^{(r-1)}$$

in the spanning set for K_n. We must prove that *every* $f \in H^2$ is the norm limit of a sequence of the form $\{f_n : f_n \in K_n\}$.

To this end, let us calculate the distance, in H^2, from f to K_n. By the Hahn-Banach distance formula, this distance is equal to

$$\sup \left\{ | \int_0^{2\pi} f\overline{g} \frac{d\theta}{2\pi} | : g \in \text{ball}(H^2), g \perp K_n \right\}.$$

By an application of the reproducing property of the k_w's,

$$K_n = (B_n H^2)^\perp,$$

where B_n is the Blaschke product with zero set T_n (including multiplicities). Thus

$$\text{dist}(f, K_n) = \sup \left\{ | \int_0^{2\pi} f\overline{hB_n} \frac{d\theta}{2\pi} | : h \in \text{ball}(H^2) \right\}.$$

To show this tends to zero as $n \to \infty$, it suffices to show

(6.1.6)
$$\lim_{n\to\infty} B_n(z) = 0 \text{ for all } z \in \mathbb{D},$$

since then, for any sequence $\{h_n\} \subset \text{ball}(H^2)$, $h_n B_n$ tends to zero in the weak topology of H^2 (see [**31**, p. 272]). To prove eq.(6.1.6), we first prove

(6.1.7)
$$\lim_{n\to\infty} B_n(0) = 0.$$

Indeed,

$$|B_n(0)| = \prod_{j=1}^{N(n)} |a_{n,j}|.$$

For any finite Blaschke product B with zeros $\{z_j\}$ (repeated according to multiplicity), let $s(B) = \sum_j (1 - |z_j|)$. With $t_j := 1 - |a_{n,j}|$, note that

$$|B_n(0)| = \prod_{j=1}^{N(n)} (1 - t_j) \leq \prod_{j=1}^{N(n)} e^{-t_j} = e^{-s(B_n)} = e^{-p_n} \to 0 \text{ as } n \to \infty.$$

This proves eq.(6.1.7).

For general $b \in \mathbb{D}$, let $g(z) := (z - b)/(1 - \overline{b}z)$. A calculation shows that

(6.1.8)
$$\frac{1 - |g(z)|^2}{1 - |z|^2} = \frac{1 - |b|^2}{|1 - \overline{b}z|^2}.$$

The composition $C_n := B_n \circ g^{-1}$, where g^{-1} is the Möbius transformation inverse of g, is a Blaschke product whose zeros are $\{g(a_{n,j}) : j = 1, \ldots, N(n)\}$. Thus,

$$(6.1.9) \qquad s(C_n) = \sum_{j=1}^{N(n)} (1 - |g(a_{n,j})|) \geq \frac{1}{2} \sum_{j=1}^{N(n)} (1 - |g(a_{n,j})|^2)$$

and, using eq.(6.1.8),

$$1 - |g(a_{n,j})|^2 = (1 - |a_{n,j}|^2) \frac{1 - |b|^2}{|1 - \bar{b}a_{n,j}|^2} \geq \frac{1 - |b|}{1 + |b|} (1 - |a_{n,j}|^2).$$

Hence, from eq.(6.1.9),

$$s(C_n) \geq \frac{1}{2} \frac{1 - |b|}{1 + |b|} s(B_n) \to \infty \quad \text{as } n \to \infty.$$

So, by eq.(6.1.7),

$$0 = \lim_{n \to \infty} C_n(0) = \lim_{n \to \infty} B_n(b),$$

completing the proof of eq.(6.1.6) and hence the proof of the theorem. $\square$

REMARK 6.1.10.   (1) There are various generalizations of this result. For example, if for each $n \in \mathbb{N}$, P_n is a finite sequence of points from $\mathbb{T}^c$ and

$$(6.1.11) \qquad \lim_{n \to \infty} \sum_{z \in P_n \cap \mathbb{D}} (1 - |z|) = \infty \quad \text{and} \quad \lim_{n \to \infty} \sum_{z \in P_n \cap \mathbb{D}_e} (1 - |z|^{-1}) = \infty,$$

then every $f \in L^p$ ($0 < p < \infty$), can be approximated (in L^p) by a sequence of rational functions whose poles lie in P_n (with appropriate order poles depending on the number of times a point appears in the sequence P_n). For $1 < p < \infty$, this result follows from generalizing the above H^2 result to H^p, which is not too difficult using the standard (H^p, H^q) duality represented by the integral pairing

$$\int_0^{2\pi} f\bar{g} \frac{d\theta}{2\pi},$$

where $1/p + 1/q = 1$ (and the argument used above), and then employing the Riesz projection theorem [**52**, p. 54] (any L^p function can be written as $f + \bar{g}$, where $f, g \in H^p$). As to be expected, the $p \leq 1$ case is more subtle. There is a Walsh type result in which L^p is replaced with $L^p(w, \mathbb{T})$, where w is some reasonable weight. A nice summary of this can be found in [**86**].

(2) When eq.(6.1.11) fails, there are Tumarkin type theorems to determine exactly which functions $f \in L^p(w, \mathbb{T})$ can be represented as

$$f = \lim_{n \to \infty} f_n, \quad f_n \in R_n,$$

see [**86**, **140**] .

(3) There are also Tumarkin type theorems in other spaces of analytic functions (see Chapter 8, § 8.7 of these notes) [**73**] [**103**, p. 39].

6.2. Definition and basic examples

We now make the notion of 'pseudocontinuation', which is the essence of Tumarkin's result, more precise.

DEFINITION 6.2.1. Let Ω be a region (an open connected set) in the extended exterior disk $\mathbb{D}_e$ which shares a non-degenerate boundary arc I with the unit disk $\mathbb{D}$. We say that $T_f \in \mathfrak{M}(\Omega)$ is a 'pseudocontinuation' of $f \in \mathfrak{M}(\mathbb{D})$ across I if the non-tangential limits of T_f and f exist and are equal almost everywhere on I.

REMARK 6.2.2.   (1) From the Lusin-Privalov uniqueness theorem (Theorem 2.2.2) it follows that if $f \in \mathfrak{M}(\mathbb{D})$ has a pseudocontinuation $T_f \in \mathfrak{M}(\Omega)$ across I, it is unique.
  (2) Here is a good place to point out that the uniqueness of a pseudocontinuation comes from the use of a *non-tangential* limit in the definition of a pseudocontinuation as opposed to say a *radial* limit, *i.e.*,

$$\lim_{r \to 1^-} f(re^{i\theta}) = \lim_{r \to 1^+} T_f(re^{i\theta}) \quad \text{a.e.}$$

One can construct a non-trivial analytic function F on the exterior disk for which

$$\lim_{r \to 1^+} F(re^{i\theta}) = 0 \quad \text{a.e.} \quad [1]$$

The function $f \equiv 0$ on the disk will then have two 'pseudocontinuations' (if one replaces 'non-tangential limits' with 'radial limits' in the definition of pseudocontinuation), the zero function (on the exterior disk) and the non-trivial function F. To avoid this pathology, we use non-tangential limits in our definition of pseudocontinuation.
  (3) It follows from the Lusin-Privalov uniqueness theorem that pseudocontinuation is compatible with analytic continuation in the sense that if $f \in \mathfrak{M}(\mathbb{D})$ has a pseudocontinuation $T_f \in \mathfrak{M}(\Omega)$ across I and f has an analytic continuation (also denoted by f) to a neighborhood U of a boundary point $e^{i\theta} \in I$, then $T_f = f$ on $U \cap \Omega$.

EXAMPLE 6.2.3. *Isolated winding points*: We say a function $f \in \mathfrak{H}(\mathbb{D})$ has an 'isolated winding point' at $\zeta \in \mathbb{T}$ if there is a punctured disk $G_\zeta = \{z : 0 < |z - \zeta| < a\}$ such that (i) f is analytically continuable (without singularities) along every path $\gamma \subset G_\zeta$ originating in $\mathbb{D}$ and (ii) there is some closed path $\gamma \subset G_\zeta$, originating in $\mathbb{D}$ along which the continuation of f does not return to its original branch. For example, $f(z) = \log(1 - z)$ has an isolated winding point at $z = 1$.

Because of the requirement of compatibility of pseudocontinuation with analytic continuation, functions with isolated winding points, do not have pseudocontinuations across any arc of $\mathbb{T}$, containing the winding point, to any contiguous domain. This can be seen as follows: Consider the annular domain R bounded by circles of radii t and $3t$ centered at ζ (the isolated winding point). We assume t is so small that f extends analytically throughout the outer disk (as a multiple-valued

---

[1]For example, Bagemihl and Seidel [16] [36, p. 163] show that given *any* continuous function g on $\mathbb{D}$ and *any* set E of first category in $\mathbb{T}$, there is an $f \in \mathfrak{H}(\mathbb{D})$ such that $f(re^{it}) - g(re^{it}) \to 0$ as $r \to 1$ for all $e^{it} \in E$.

function), except for the winding point at ζ. Suppose now that f has a pseudo-continuation $T_f \in \mathfrak{M}(\Omega)$ across an open arc I containing ζ, where $\Omega \subset \mathbb{D}_e$ with $\Omega^- \cap \mathbb{T} = I$. Assume here that $\Omega \subset \{|z - \zeta| < 3t\}$. Let K denote the circle of radius $2t$ centered at ζ, and suppose P and Q are points on K with $P \in \mathbb{D}$ and $Q \in \mathbb{D}_e$. If we now apply the compatibility of analytic continuation with pseudocontinuation (see above) to the two possible analytic continuations of f from P to Q along arcs of K, the values of both these continuations at Q are required to be equal to the same number, $T_f(Q)$. This contradicts the assumed branching of f.

As we will make more precise in Chapter 9, functions with isolated winding points do not have 'generalized analytic continuations' of any sort.

EXAMPLE 6.2.4. *Inner functions*:

(1) Let $\{a_n\}$ be a sequence of points in $\mathbb{D} \setminus \{0\}$ (repeated according to multiplicity) such that

$$\sum_n (1 - |a_n|) < \infty.$$

The 'Blaschke product'

$$b(z) := \prod_n \frac{|a_n|}{a_n} \frac{a_n - z}{1 - \overline{a_n}z}$$

converges uniformly on compact subsets of $\mathbb{D}$ to define a bounded analytic function on $\mathbb{D}$ whose boundary function $b(e^{i\theta})$ satisfies $|b(e^{i\theta})| = 1$ for almost every θ. Furthermore, the product converges uniformly on the compact subsets of $\mathbb{D}_e$, which do not intersect $\{1/\overline{a_n}\}^-$, to a meromorphic function on $\mathbb{D}_e$ with poles at the points $1/\overline{a_n}$. Directly from the definition of b it follows that

$$b(z) = \frac{1}{\overline{b(1/\overline{z})}} \quad \text{on } \mathbb{D}_e \setminus \{z : b(1/\overline{z}) = 0\}$$

which, using the fact that $|b(e^{i\theta})| = 1$ for almost every θ, shows that $b|\mathbb{D}_e$ is a pseudocontinuation of $b|\mathbb{D}$ across $\mathbb{T}$. If we allow $\{a_n\}$, the zeros of b, to accumulate on all of $\mathbb{T}$, then b will have a pseudocontinuation across $\mathbb{T}$ but not an analytic continuation across any point of $\mathbb{T}$.

(2) Let μ be a positive singular measure on $\mathbb{T}$ (with respect to Lebesgue measure) and define the 'singular inner function'

$$s_\mu(z) := \exp\left(-\int_{\mathbb{T}} \frac{\zeta + z}{\zeta - z} d\mu(\zeta)\right).$$

One can check (see reference below) that s_μ is a bounded analytic function on $\mathbb{D}$ whose boundary function $s_\mu(e^{i\theta})$ satisfies $|s_\mu(e^{i\theta})| = 1$ for almost every θ. By differentiating under the integral, it is easy to verify that s_μ defines an analytic function on the complement of the support of μ. This domain of analyticity is maximal since one can show that neither s_μ, nor even $|s_\mu|$, can be continuously extended to any point in the support of μ. Again, it follows from the definition of s_μ that

$$s_\mu(z) = \frac{1}{\overline{s_\mu(1/\overline{z})}} \quad \text{on } \mathbb{D}_e.$$

So, using the fact that $|s_\mu(e^{i\theta})| = 1$ for almost every θ, $s_\mu|\mathbb{D}_e$ is a pseudocontinuation of $s_\mu|\mathbb{D}$ across $\mathbb{T}$. If the support of μ is equal to $\mathbb{T}$, then s_μ will have no analytic continuation across any point of $\mathbb{T}$. We refer the reader to [**60**, pp. 75 - 76] for the proofs of the above results.

(3) If f is any 'inner function', that is to say, a bounded analytic function on $\mathbb{D}$ whose boundary function $f(e^{i\theta})$ satisfies $|f(e^{i\theta})| = 1$ a.e. θ then, by standard theory (see Chapter 2), $f = z^m b s_\mu$ for some $m = 0, 1, 2, \ldots$, Blaschke product b, and singular inner function s_μ. Thus f has a pseudocontinuation across $\mathbb{T}$ given by

$$T_f(z) := \frac{1}{\overline{f(1/\bar{z})}}.$$

EXAMPLE 6.2.5. *Cauchy transforms*: If μ is a finite singular measure on $\mathbb{T}$ (with respect to Lebesgue measure on $\mathbb{T}$), then from Smirnov's theorem (Theorem 3.2.6) and Fatou's theorem (Theorem 3.2.9), the component functions, $f_\mu|\mathbb{D}$ and $f_\mu|\mathbb{D}_e$, of the Cauchy transform

$$f_\mu(z) = \int \frac{1}{1 - ze^{-it}} d\mu(e^{it})$$

are pseudocontinuations of each other across $\mathbb{T}$. Recall from Chapter 3 that the Poincaré example

$$f(z) = \sum_{n=1}^\infty \frac{c_n}{1 - e^{-i\theta_n} z},$$

where $\{c_n\}$ is an absolutely summable sequence of complex numbers, is a special case of this. Again, note (Theorem 3.1.1) that if $\{e^{i\theta_n}\}$ is a dense set in $\mathbb{T}$, then f_μ need not have an analytic continuation across any point $e^{i\theta}$.

EXAMPLE 6.2.6. *Certain Borel series*: From Proposition 4.2.19, the component functions $f|\mathbb{D}$ and $f|\mathbb{D}_e$ of the Borel series

$$f(z) := \sum_{n=1}^\infty \frac{c_n}{z - z_n},$$

where $\{c_n\}$ is an absolutely summable sequence and $\{z_n\} \subset \mathbb{D}$ satisfies the Blaschke condition $\sum(1 - |z_n|) < \infty$, are pseudocontinuations of each other. As we shall see in eq.(6.6.6), not all Borel series have pseudocontinuations.

EXAMPLE 6.2.7. *Gap series*: Functions such as

$$f(z) = \sum_{n=0}^\infty z^{2^n} 2^{-n}, \quad |z| < 1,$$

which have radius of convergence equal to one, do not have analytic continuations across any point of the unit circle (see Theorem 6.9.2 and Theorem 6.9.4). As we shall see in § 6.9 of this chapter, such functions (as well as a host of other gap series) do not have pseudocontinuations either.

6.3. Cyclic vectors for the backward shift on H^2

Shift operators on various function spaces play an important role in operator theory. They often serve as 'models' for certain types of operators (normal, subnormal, etc.) and understanding the properties of shift operators, especially their invariant and co-invariant subspaces, leads to a greater understanding of general operators of certain types (see [103]). Perhaps the simplest type of shift operator on a Hilbert space of analytic functions is the 'unilateral shift operator'

$$S : H^2 \to H^2, \quad (Sf)(z) = zf(z).$$

One of the most studied aspects of the unilateral shift, and the aspect which has been generalized to a variety of settings, is its invariant subspace structure. For any inner function ϕ, the subspace ϕH^2 is a closed subspace of H^2 and is clearly 'invariant' for S, that is to say,

$$S(\phi H^2) \subset \phi H^2.$$

Moreover, a well-known theorem of Beurling [21] says these are all of them.

THEOREM 6.3.1 (Beurling). *Let $\mathcal{M} \neq (0)$ be a closed S-invariant subspace of H^2. Then $\mathcal{M} = \phi H^2$ for some inner function ϕ.*

One can also ask about the 'cyclic vectors' for S, that is to say, the vectors $f \in H^2$ which are not contained in any proper invariant subspace of H^2, equivalently, those f for which

$$\bigvee \{ S^n f : n = 0, 1, 2, \dots \} = H^2.$$

Such cyclic vectors f are well understood by another result of Beurling.

THEOREM 6.3.2 (Beurling). *A function $f \in H^2$ is cyclic for S if and only if f is an 'outer function', that is to say,*

$$\log |f(0)| = \int_0^{2\pi} \log |f| \frac{d\theta}{2\pi}.$$

The important operator to study here, in that it is close to our central theme of continuation properties of analytic functions, is the adjoint of S. With the usual Hilbert space pairing

$$\langle f, g \rangle = \int_0^{2\pi} f \overline{g} \frac{d\theta}{2\pi},$$

it is easy to check that the adjoint, S^*, of S is given by

$$(S^* f)(z) = \frac{f(z) - f(0)}{z} = a_1 + a_2 z + a_3 z^2 + \cdots$$

and is therefore called the 'backward shift operator'. From elementary functional analysis and Beurling's theorem (Theorem 6.3.1), every S^*-invariant subspace is of the form $(\phi H^2)^\perp$, where ϕ is an inner function. Though there is characterization of $(\phi H^2)^\perp$, which we will get to shortly, we first focus our attention on the corresponding 'cyclic vector' problem for S^*. Here, a function $f \in H^2$ is 'non-cyclic' for S^* if

$$\bigvee \{ S^{*n} f : n = 0, 1, 2, \dots \} \neq H^2.$$

From what was said above, this is so if and only if $f \perp \phi H^2$ for some inner ϕ. As it turns out, the concept of a pseudocontinuation gives us a more satisfactory description of these non-cyclic vectors.

DEFINITION 6.3.3.   (1) Define

$$\mathfrak{N}(\mathbb{D}_e) := \left\{ \frac{G}{H} : G, H \in H^\infty(\mathbb{D}_e) \right\}$$

to be the 'functions of bounded type' (in Nevanlinna's sense) on $\mathbb{D}_e$, see Chapter 2.

(2) Set $PCBT$ to be the class of $f \in H^2$ which have a pseudocontinuation across $\mathbb{T}$ to a function $T_f \in \mathfrak{N}(\mathbb{D}_e)$.

The main theorem characterizing the cyclic vectors for S^* on H^2 is the following.

THEOREM 6.3.4 (Douglas-Shapiro-Shields [51]). *A function $f \in H^2$ is non-cyclic for S^* if and only if $f \in PCBT$.*

REMARK 6.3.5.   (1) Note that Tumarkin's result (Theorem 6.1.3) gives an equivalent condition for cyclicity.

(2) From the examples in the previous section, inner functions are non-cyclic vectors while H^2 functions with an isolated winding point (for example, $\log(1-z)$) and certain H^2 gap series are cyclic vectors.

(3) The point at infinity is important. For example, the function $f = e^z$, although entire, does not belong to $PCBT$ due to its essential singularity at infinity.

It is worth mentioning that if f is a non-cyclic vector and g is any non-zero vector which annihilates the S^*-invariant subspace generated by f, that is to say, $\langle S^{*n} f, g \rangle = 0$ for all $n = 0, 1, 2, \ldots$, the function

$$(6.3.6) \qquad C_{f,g}(\lambda) := \left\langle \frac{zf}{z-\lambda}, g \right\rangle \Big/ \left\langle \frac{\lambda}{z-\lambda}, g \right\rangle, \quad \lambda \in \mathbb{D}_e,$$

is a formula for the pseudocontinuation of f. One can see this by noting a few things. First, writing the inner products in integral form shows that $C_{f,g}$ is the quotient of two Cauchy transforms (Definition 3.2.5) and so, from Smirnov's theorem (Theorem 3.2.6), $C_{f,g}$ is a function of bounded type. Second, a simple power series computation shows that

$$(6.3.7) \qquad \left\langle \frac{\lambda}{z-\lambda}, g \right\rangle = \lambda \int_0^{2\pi} \frac{1}{e^{it} - \lambda} \overline{g} \frac{dt}{2\pi} = -\overline{g}(1/\overline{\lambda}), \quad |\lambda| > 1.$$

Third, for $|\lambda| < 1$, the expression

$$\left\langle \frac{zf}{z-\lambda}, g \right\rangle = \int_0^{2\pi} \frac{e^{it} f \overline{g}}{e^{it} - \lambda} \frac{dt}{2\pi}$$

is equal to

$$\sum_{n=0}^{\infty} \lambda^n \int_0^{2\pi} e^{-int} f\overline{g} \frac{dt}{2\pi} = \sum_{n=0}^{\infty} \lambda^n \langle S^{*n} f, g \rangle = 0,$$

since g annihilates the S^*-invariant subspace generated by f. By Fatou's theorem (Theorem 3.2.9) and the previous equation,

$$\lim_{r \to 1^-} \int_0^{2\pi} \frac{e^{it} f \bar{g}}{e^{it} - (e^{i\theta}/r)} \frac{dt}{2\pi} = -f(e^{i\theta}) \bar{g}(e^{i\theta}) \quad \text{a.e. } \theta.$$

Combining this with eq.(6.3.7) and the definition of $C_{f,g}$, one can see that the non-tangential boundary values of $C_{f,g}$ (from outside the disk) are equal to those of f (from inside the disk) almost everywhere. Hence $C_{f,g}$ is a pseudocontinuation of f for each non-trivial g in the annihilator.

From the Lusin-Privalov uniqueness theorem (Theorem 2.2.2), observe that $C_f := C_{f,g}$ is independent of the annihilating g. This independence will be important later (Proposition 8.4.13) when studying the spectral properties of the operator $S^*|\mathcal{M}$, where $\mathcal{M}$ is an S^*-invariant subspace of H^2.

Although the concept of a pseudocontinuation gives us a complete answer to the cyclic vector problem for the backward shift on H^2, it is rather unsatisfactory in the sense that it is nearly impossible to check whether or not a generic function from H^2 has a pseudocontinuation of bounded type. At present, there does not seem to be any reasonable condition for cyclicity for S^* as there is for the forward shift S (f is cyclic for S if and only if f is an outer function). There are some results which attempt to understand the cyclic vector problem for S^* in a more tangible way. We list a few here.

(1) The first set of results, found in the Douglas, Shapiro, Shields paper [51], discuss when a function $f \in H^2$ has an analytic continuation across a portion of the circle.
   (a) We know from before that if $f \in H^2$ has an isolated winding point at $e^{i\theta} \in \mathbb{T}$, then f is cyclic.
   (b) If $f \in H^2$ is analytic on $R\mathbb{D}$ where $R > 1$, then f is either cyclic or is a rational function (and hence not cyclic).
(2) There is some information which can be obtained about cyclic vectors from the growth of its Taylor coefficients. For example, H. S. Shapiro [127], in perhaps one of the earliest published results about cyclic vectors, showed that if

$$f = \sum_{n=0}^{\infty} a_n z^n, \quad \text{with} \quad \lim_{n \to \infty} \sqrt[n]{|a_n|} = 0$$

and f is not a polynomial, then f is cyclic for H^2. Kriete [92] was able to obtain further information along these lines. He showed that if $f = \sum_n a_n z^n \in H^2$ is a non-cyclic vector and not a polynomial, then there is an $r > 0$ and $k = 0, 1, \ldots$ such that

$$\sum_{n=k+j}^{\infty} |a_n|^2 \geq r^j \sum_{n=k}^{\infty} |a_n|^2 \quad \text{for all } j = 1, 2, \ldots$$

Unfortunately, this condition does not characterize the cyclic vectors. In fact, given $\{d_j\} \subset (0, 1)$ with $d_j \to 0$, there is a bounded cyclic outer function $f = \sum_n a_n z^n$ such that

$$\sum_{n=j}^{\infty} |a_n|^2 \geq d_j \|f\|_\infty^2.$$

(3) The next set of results relate cyclicity with the modulus.
  (a) If $f \in H^2$ is non-constant and

$$\int \log |\Re f(e^{i\theta})| d\theta = -\infty,$$

  then f is cyclic [51].
  (b) (Kriete [92]) If $f \in H^2$ is non-cyclic and $a > 0$. Then either

$$\int \log \big| \, a - |f(e^{i\theta})| \, \big| \, d\theta > -\infty,$$

  or $|f| = a$ almost everywhere.
  (c) (Kriete [92]) If $f, g \in H^2$ are non-cyclic, then either

$$\int \log \big| \, |f(e^{i\theta})| - |g(e^{i\theta})| \, \big| \, d\theta > -\infty,$$

  or $|f| = |g|$ almost everywhere.

There is also a series of results [50, 51] which explore the properties of the sets $\mathcal{N}$ (the non-cyclic vectors) and $\mathcal{C}$ (the cyclic vectors).

THEOREM 6.3.8. (1) $H^2 = \mathcal{C} + \mathcal{C}$.
  (2) $\mathcal{N}$ is a dense F_σ set of first category in H^2.
  (3) $\mathcal{C}$ is a dense G_δ set in H^2.
  (4) $\mathcal{N} + \mathcal{N} \subset \mathcal{N}$.
  (5) $\mathcal{N} + \mathcal{C} \subset \mathcal{C}$.
  (6) $(\mathcal{N} \cdot \mathcal{N}) \cap H^2 \subset \mathcal{N}$.
  (7) $(\mathcal{N} \cdot \mathcal{C}) \cap H^2 \subset \mathcal{C}$.
  (8) If $f \in \mathcal{C}$ and $1/f \in H^2$, then $1/f \in \mathcal{C}$.

The description of $(\phi H^2)^\perp$, and hence the complete description of the S^*-invariant subspaces of H^2, is the following.

THEOREM 6.3.9 (Douglas-Shapiro-Shields [51]). *A function $f \in H^2$ belongs to $(\phi H^2)^\perp$, where ϕ is an inner function, if and only if f/ϕ has a pseudocontinuation $T_{f/\phi}$ across $\mathbb{T}$ such that $T_{f/\phi} \in H^2(\mathbb{D}_e)$ and $T_{f/\phi}(\infty) = 0$. Moreover,*

$$(\phi H^2)^\perp = \bigvee \{ S^{*n} \phi : n = 1, 2, 3, \dots \},$$

that is to say, $(\phi H^2)^\perp$ is a 'cyclic subspace' generated by the vector $S^* \phi$.

REMARK 6.3.10. (1) The characterization of the non-cyclic vectors for the backward shift (Theorem 6.3.4) can be generalized to H^p ($1 < p < \infty$) with nearly the same proof by using the standard (H^p, H^q) duality. For $p = 1$, the result is the same except that the duality theory is more delicate since it involves the functions of bounded mean oscillation. For $0 < p < 1$, Theorem 6.3.4 is still true except that the proof, due to Aleksandrov [4], is very different and much more complicated (see below).
  (2) The description of the backward shift invariant subspaces of H^p ($1 \le p < \infty$) is similar to that of H^2. The case $p = 1$ is a bit tricky and was done by Aleksandrov in [4]. The full details, in English, can be found in [35, p. 101]. In [5], Aleksandrov characterizes the backward shift invariant

subspaces of many other spaces of analytic functions which are related to the Hardy spaces, such as $VMOA$, $BMOA$, and $L^1/\overline{H_0^1}$.

(3) The description of the backward shift invariant subspaces of H^p $(0 < p < 1)$, again due to Aleksandrov [4], is more complicated and the proof involves many advanced tools from analysis such as Coifman's atomic decomposition. One of the main difficulties is that H^p $(0 < p < 1)$ is no longer a locally convex space and hence the Hahn-Banach separation theorem, a key tool in the analysis of the $p \geq 1$ case, is no longer available. For example, there are non-trivial backward shift invariant subspaces of H^p $(0 < p < 1)$, for example,

$$\bigvee \left\{ \frac{1}{1 - e^{-i\theta}z} : 0 \leq \theta < 2\pi \right\},$$

whose annihilator (see [53] for the dual of H^p $(0 < p < 1)$) is the zero subspace. A complete account of all this can be found in [35]. A discussion of the backward shift operator on the Smirnov class N^+ can be found in another paper of Aleksandrov [6].

(4) The backward shift operator on H^2 enjoys an important place in the study of general linear operators on Hilbert spaces. Important work beginning with Rota [119] and continuing with de Branges and Rovnyak [40] and then Foiaş [57] show, for example, that any operator T on a separable Hilbert space satisfying the two conditions (i) $\|T\| \leq 1$ and (ii) $\|T^n x\| \to 0$ as $n \to \infty$ for all x, is unitarily equivalent to a direct sum of backward shifts restricted to an invariant subspace (of the direct sum). The language used in the literature says that T is unitarily equivalent to 'part' of a direct sum of backward shifts.

6.4. The Hardy space of a multiply connected domain

Pseudocontinuations also play an important role in examining the various types of invariant subspaces for the operator $M_z f = zf$ for the Hardy space of a multiply connected domain [10, 76, 120, 121, 147]. Without getting into too much detail here which would take us too far afield, we describe what happens in the easiest of multiply connected domains, an annulus $G = \{1 < |z| < R\}$.

The Hardy space $H^2(G)$ is the space of functions $f \in \mathfrak{H}(G)$ which have finite norm

$$\|f\|^2 = \sup_{1 < r < R} \int_0^{2\pi} |f(re^{i\theta})|^2 d\theta.$$

There is an equivalent definition of $H^2(G)$ using harmonic majorants [120]. The operator $M_z f = zf$ is a continuous linear operator on $H^2(G)$ and one can seek a characterization of the invariant subspaces $\mathcal{M}$ of M_z. There are two basic types:

'fully invariant': $z\mathcal{M} \subset \mathcal{M}$ and $z^{-1}\mathcal{M} \subset \mathcal{M}$

'simply invariant': $z\mathcal{M} \subset \mathcal{M}$ but $z^{-1}\mathcal{M} \not\subset \mathcal{M}$.

In a generalization of Beurling's theorem, Sarason [121] characterized the fully invariant subspaces of $H^2(G)$.

THEOREM 6.4.1 (Sarason). *Let $\mathcal{M} \subset H^2(G)$ be a fully invariant subspace. Then there is a G-inner function ϕ such that $\mathcal{M} = \phi H^2(G)$.*

A bounded analytic function on G is 'G-inner' if the non-tangential boundary values of $|\phi|$ on each boundary contour of G are almost everywhere equal to a constant. Note that this definition makes the zero function a G-inner function.

The characterization of the simply invariant subspaces is where pseudocontinuations come in. This following result of Royden [120], for which we shall provide a proof, characterizes certain simply invariant subspaces of $H^2(G)$.

THEOREM 6.4.2 (Royden). *Suppose that $\mathcal{M}$ is a simply invariant subspace of $H^2(G)$ which contains the constant function 1. Then there is an inner function ϕ on $\mathbb{D}$ such that $\mathcal{M}$ is the space of functions $f \in H^2(G)$ with the property that f has a pseudocontinuation across $\mathbb{T}$ to a function $T_f \in \mathfrak{N}(\mathbb{D})$ for which $\phi T_f \in H^2(\mathbb{D})$.*

PROOF. For any $f \in H^2(G)$ note that

$$f = \sum_{k=1}^{\infty} a_k z^k + \sum_{k=0}^{\infty} b_k/z^k = f_+ + f_-.$$

Note that $f_+ \in H^2(R\mathbb{D})$ and $f_- \in H^2(\mathbb{D}_e)$. Furthermore, the operator

$$S_{\infty}^* : H^2(\mathbb{D}_e) \to H^2(\mathbb{D}_e), \quad S_{\infty}^* g = z(g - g(\infty))$$

is continuous and invertible with $S_{\infty}^* U = U S^*$, where $U : H^2 \to H^2(\mathbb{D}_e)$ is given by $(Uh)(z) = h(1/z)$.

Using the above decomposition, our M_z-invariant subspace $\mathcal{M}$ can be written as $\mathcal{M} = \mathcal{M}_+ + \mathcal{M}_-$, where $\mathcal{M}_+ = \{f_+ : f \in \mathcal{M}\}$ and $\mathcal{M}_- = \{f_- : f \in \mathcal{M}\}$. Moreover, since $1 \in \mathcal{M}_-$, then $S_{\infty}^* \mathcal{M}_- \subset \mathcal{M}_-$. The subspace $U^{-1}\mathcal{M}_-$ is an S^*-invariant subspace of H^2, which (Theorem 6.3.9) takes the form m_ϕ of functions $g \in H^2$ such that g/ϕ has a pseudocontinuation across $\mathbb{T}$ to a function $T_{g/\phi} \in H^2(\mathbb{D}_e)$ with $T_{g/\phi}(\infty) = 0$. Here ϕ is an inner function on $\mathbb{D}$. We leave it to the reader to check that Um_ϕ is the set of functions $f \in H^2(\mathbb{D}_e)$ such that f has a pseudocontinuation across $\mathbb{T}$ to a function $T_f \in \mathfrak{N}(\mathbb{D})$ such that $\phi T_f \in H^2(\mathbb{D})$ and ϕT_f vanishes at the origin. The proof now follows from the fact that $\mathcal{M}_+ \subset H^2(R\mathbb{D})$. $\square$

Royden [120] went on further to show that if $\mathcal{M}$ is any simply invariant subspace and f and g are non-zero functions in $\mathcal{M}$, then f/g has a pseudocontinuation across $\mathbb{T}$ to a function of bounded type in $\mathbb{D}$. The final piece of the puzzle, at least for the annulus case, was done by Hitt [76].

THEOREM 6.4.3 (Hitt). *Let $\mathcal{M}$ be a simply invariant subspace of $H^2(G)$. Then*

$$\mathcal{M} = z^n F \Phi \mathcal{M}_\phi,$$

where $n \in \mathbb{N} \cup \{0\}$, $F(z) = O(1/z)$ for some outer function O, Φ is G-inner, ϕ is an inner function on $\mathbb{D}$, and $\mathcal{M}_\phi$ is the space of functions $f \in H^2(G)$ such that f has a pseudocontinuation across $\mathbb{T}$ to a function $T_f \in \mathfrak{M}(\mathbb{D})$ for which $\phi T_f \in H^2(\mathbb{D})$.

The above results were generalized to more complicated multiply connected domains by Royden [120], Yakubovich [147], and Aleman and Richter [10]. There is a result similar to Royden's theorem (Theorem 6.4.2) in [14] for the Bergman space of an annulus.

6.5. Walsh-Tumarkin again

The classification of the S^*-invariant subspaces of H^2 also makes connections to the Tumarkin result mentioned earlier in Theorem 6.1.3. For $|\lambda| < 1$ and

$$k_\lambda(z) := \frac{1}{1 - \overline{\lambda}z},$$

we have

$$(\overline{\lambda}I - S^*)k_\lambda = 0.$$

A computation also shows that for any $n \in \mathbb{N}$,

$$(\overline{\lambda}I - S^*)^n k_\lambda^{(j)} = 0, \quad j = 0, \ldots, n-1,$$

where

(6.5.1)
$$k_\lambda^{(j)}(z) := \frac{j!z^j}{(1 - \overline{\lambda}z)^{j+1}},$$

which means that each $k_\lambda^{(n)}$ ($\lambda \in \mathbb{D}$, $n = 0, 1, \ldots$) is a 'root vector' for S^*. In fact,

$$\ker(\overline{\lambda}I - S^*)^n = \bigvee\{k_\lambda^{(j)} : j = 0, \ldots, n-1\}.$$

The well-known Kronecker theorem of linear algebra [**78**, p. 182] says that any finite dimensional S^*-invariant subspace is the linear span of its root vectors. For an infinite dimensional invariant subspace $\mathcal{M}$, is a similar result true? Is

(6.5.2)
$$\mathcal{M} = \bigvee \{ k_\lambda^{(n)} : k_\lambda^{(n)} \in \mathcal{M} \}?$$

DEFINITION 6.5.3. Invariant subspaces $\mathcal{M}$ satisfying eq.(6.5.2) are said to admit 'spectral synthesis'.

Using the identity

$$\langle g, k_\lambda^{(n)} \rangle = g^{(n)}(\lambda), \quad g \in H^2,$$

one can see that

$$\ker(\overline{\lambda}I - S^*)^n = (b_\lambda H^2)^\perp,$$

where b_λ is the Blaschke product with an n-fold zero at λ (and no other zeros). Thus $\mathcal{M}$ admits spectral synthesis, if and only $\mathcal{M}^\perp$ is a 'zero-based' S-invariant subspace of H^2, or equivalently, $\mathcal{M}^\perp = bH^2$, where b is a Blaschke product. This means, for example, that $\mathcal{M} = (s_\mu H^2)^\perp$, where s_μ is a singular inner function, is an S^*-invariant subspace which does not admit spectral synthesis. Perhaps a suitable substitute for spectral synthesis is what has been called, in the work of Gribov and Nikol'skiĭ [**64**] [**103**] and Shimorin [**133**], 'approximate spectral synthesis'.

DEFINITION 6.5.4. An S^*-invariant subspace $\mathcal{M}$ is said to admit 'approximate spectral synthesis' if there is a sequence of S^*-invariant subspaces R_n, each consisting of the linear span of a finite number of root vectors, as in eq.(6.5.1), such that

$$\mathcal{M} = \underline{\lim} \, R_n,$$

where

$$\underline{\lim} \, R_n := \{ f \in H^2 : \lim_{n \to \infty} \, \mathrm{dist}(f, R_n) = 0 \} .$$

Using a slightly different notation than we used in our earlier discussion of the Walsh-Tumarkin results (Theorem 6.1.2 and Theorem 6.1.3), we let

$$S_n := \{z_{n,1}, z_{n,2}, \ldots, z_{n,N(n)}\}, \quad n = 1, 2, \ldots$$

denote a tableau of points in $\mathbb{D}$ and R_n be the linear span of the corresponding root vectors, that is to say

$$R_n = \bigvee \left\{ k_{z_{n,j}}^{(p)} : j = 1, \ldots, N(n), p = 0, \ldots, \text{mult}(z_{n,j}) - 1 \right\},$$

where $\text{mult}(z_{n,j})$ is the number of times the point $z_{n,j}$ appears in the sequence S_n. Clearly $\varlimsup R_n$ is an S^*-invariant subspace of H^2 since in fact, $R_n = (b_n H^2)^\perp$, where b_n is the Blaschke product with zeros (with appropriate multiplicity) at the points of S_n (and no other zeros). A result of Tumarkin [140] (see also [51, Thm. 4.2.1] [103, p. 37]) says these are all of them.

THEOREM 6.5.5 (Tumarkin). *If $\mathcal{M}$ is an S^*-invariant subspace of H^2, then there is a tableau $\{S_n\}$, as above, such that*

$$\mathcal{M} = \varlimsup R_n.$$

Moreover, for any such tableau $\{S_n\}$,

$$\varlimsup R_n = H^2 \Leftrightarrow \lim_{n \to \infty} \sum_{j=1}^{N(n)} (1 - |z_{n,j}|) = \infty.$$

We will examine this approximate spectral synthesis theorem for other spaces of analytic functions in Chapter 8, § 8.7.

6.6. Other spaces of analytic functions

Pseudocontinuations also come into play when looking at the backward shift operator on some other classical spaces of analytic functions. For example, the backward shift operator

$$Bf = \frac{f - f(0)}{z}$$

is continuous on the Bergman space L_a^2 (see Chapter 2 for a definition and some basic properties) [2]. The result here is the following.

THEOREM 6.6.1 (Richter-Sundberg [114]). *If f is non-cyclic for the backward shift B on L_a^2, then f is of bounded type in $\mathbb{D}$ and has a pseudocontinuation across $\mathbb{T}$ to a function T_f of bounded type in $\mathbb{D}_e$.*

REMARK 6.6.2. The fact that a non-cyclic vector must be of bounded type is quite a restriction for Bergman functions since, unlike functions in the Hardy space, Bergman functions need not be of bounded type. In fact, as we have seen in Chapter 2, Bergman functions need not have non-tangential limits on any set of positive measure in $\mathbb{T}$. By the above theorem, such poorly behaved functions would be cyclic vectors.

---

[2]We are using the notation B here, instead of S^* (as in the Hardy space setting) since the adjoint (in the Hilbert space sense of L_a^2) of S is a Bergman-Toeplitz operator with symbol $\bar{z}$. However, in the Cauchy duality, which identifies the dual of L_a^2 with the classical Dirichlet space $\mathcal{D}$, the adjoint of S is indeed B.

This cyclic vector result was first discovered by Richter and Sundberg [**114**] for the L_a^2 setting and generalized to the weighted L^p Bergman space setting in [**14**] (see also [**35**, Thm. 5.4.5]) and then to a more general Bergman-type space in [**11**]. As a consequence of this result, functions which have isolated branch points and H^2 functions which are certain types of gap series, with radius of convergence equal to one (see Theorem 6.9.7), are cyclic vectors for the backward shift on the Bergman space.

We mention that the converse to Theorem 6.6.1 is false. There are, in fact, inner functions, which certainly belong to $PCBT$ and hence are non-cyclic for the backward shift on H^2, which are cyclic for the backward shift on L_a^2. One can see this by noting that the dual of L_a^2 can be identified with the classical 'Dirichlet space' $\mathcal{D}$ (see Chapter 2) by means of the Cauchy pairing

$$\langle f, g \rangle := \lim_{r \to 1^-} \int_0^{2\pi} f(re^{i\theta}) \overline{g}(e^{i\theta}) \frac{d\theta}{2\pi}, \quad f \in L_a^2,\ g \in \mathcal{D}.$$

If ϕ is an inner function, then, by the Hahn-Banach theorem, ϕ is non-cyclic for the backward shift on L_a^2 if and only if there is a non-trivial $g \in \mathcal{D}$ for which

$$\langle B^n \phi, g \rangle = 0, \quad n = 0, 1, 2, \ldots.$$

It is routine to check that

$$0 = \langle B^n \phi, g \rangle = \langle \phi, S^n g \rangle = \int_0^{2\pi} \phi \overline{g} e^{-in\theta} \frac{d\theta}{2\pi} \text{ for all } n = 0, 1, 2, \ldots$$

and so, by the F. and M. Riesz theorem (Theorem 2.2.4), $g/\phi \in H^2$. In summary, we have shown that an inner function is non-cyclic for B on L_a^2 if and only if ϕ is the inner divisor of some non-trivial Dirichlet function. One can certainly create an inner function which is not the divisor of any Dirichlet function (by making it have too many zeros [**132**]). This inner function will belong to $PCBT$ (as all inner function do) but will be a cyclic vector for the backward shift on L_a^2. A complete description of the cyclic vectors for the Bergman space remains an open problem.

QUESTION 6.6.3. What are the cyclic vectors for the backward shift on the Bergman space?

Actually, the above question could be posed for the more general Bergman space L_a^p ($0 < p < \infty$), of analytic functions f on $\mathbb{D}$ for which

$$\int_{\mathbb{D}} |f|^p dA < \infty$$

or for the weighted space $L_a^p(w)$ of analytic f for which

$$\int_{\mathbb{D}} |f|^p w(|z|) dA < \infty,$$

where $w : [0, 1) \to \mathbb{R}_+$ is some suitable weight function. As mentioned earlier, a version of the Richter-Sundberg theorem holds for $L_a^p(w)$ ($1 \le p < \infty$) [**11**]: if $f \in L_a^p(w)$ is non-cyclic, then f is of bounded type and has a pseudocontinuation of bounded type. By using a new type of continuation outlined in Chapter 8, we can prove some other useful facts about the cyclic vectors for L_a^2. Also, we mention here that unlike the Hardy space, where $f \in H^p$ ($0 < p < \infty$) is non-cyclic if and only if $f \in PCBT$, cyclicity in the Bergman spaces L_a^p depends on the index p. Indeed

[14] [35, p. 108], there are inner functions which are cyclic for L_a^p $(1 < p < 2)$, but non-cyclic for L_a^2.

In examining the cyclic vectors for L_a^p $(1 \leq p < \infty)$, one uses duality theory and especially the Hahn-Banach separation theorem. When $0 < p < 1$, the Bergman space L_a^p is no longer a locally convex space and the Hahn-Banach separation theorem does indeed fail. In the H^p setting, Aleksandrov [4] (see also [35]) was able to overcome this difficulty and characterize the cyclic vectors for H^p $(0 < p < 1)$: $f \in H^p$ is non-cyclic if and only if $f \in PCBT$. As mentioned earlier, his proof is very complicated and depends on many advanced tools from analysis. We suspect the same level of complexity will be encountered when attempting to characterize the B-invariant subspaces of the Bergman space L_a^p $(0 < p < 1)$.

Classifying the cyclic vectors for the backward shift B on the classical Dirichlet space $\mathcal{D}$ is an even harder problem. Certainly any Dirichlet function f which also belongs to $PCBT$ is non-cyclic. Indeed, first note that

$$\mathcal{D} \not\subset \operatorname{span}_{H^2}\{B^n f : n = 0, 1, 2, \ldots\}.$$

Otherwise, using the density of $\mathcal{D}$ in H^2 (both spaces contain the polynomials),

$$H^2 = \operatorname{span}_{H^2}\{B^n f : n = 0, 1, 2, \ldots\}$$

contradicting the fact that $f \in PCBT$. Thus there is a $g \in \mathcal{D}$ and a $\delta > 0$ such that $\|p(B)f - g\|_{H^2} > \delta$ for every analytic polynomial p. But

$$\|p(B)f - g\|_{\mathcal{D}} \geq \|p(B)f - g\|_{H^2} > \delta$$

which shows that f is non-cyclic.

The converse fails miserably. In fact, one can create a non-cyclic vector for $\mathcal{D}$ which does not have a pseudocontinuation across any arc of $\mathbb{T}$. The proof involves some interesting concepts from analysis and is worth outlining here. The details can be found in [11, Thm. 6.2]. First recall from Chapter 2 that the dual of $\mathcal{D}$ can be identified with the Bergman space L_a^2 by means of the Cauchy pairing

$$\langle f, g \rangle := \lim_{r \to 1^-} \int_0^{2\pi} f(e^{i\theta})\overline{g}(re^{i\theta})\frac{d\theta}{2\pi}, \quad f \in \mathcal{D}, g \in L_a^2.$$

If $A = \{a_n\}$ is a sequence of points in $\mathbb{D}$, then

$$\mathcal{M}(A) := \bigvee \left\{ \frac{1}{1 - \overline{a_n}z} : a_n \in A \right\}$$

is a B-invariant subspace of $\mathcal{D}$. Moreover, for any $g \in L_a^2$,

$$\left\langle \frac{1}{1 - \overline{a_n}z}, g \right\rangle = \overline{g}(a_n).$$

Hence $\mathcal{M}(A) \neq \mathcal{D}$ if and only if A is the zero set of some non-trivial Bergman function. Now choose A to not only be the zero set of a non-trivial Bergman function but to also have the property that for almost every point $\zeta \in \mathbb{T}$ there is a subsequence of A converging to ζ non-tangentially [3]. We pause to give a proof of the existence of such zero sets.

PROPOSITION 6.6.4. *Let M be any increasing function on $\mathbb{R}_+$ with $M(x) \to \infty$ as $x \to \infty$. Then, there exists a non-trivial $f \in \mathfrak{H}(\mathbb{D})$ with*

---

[3]One recalls from Theorem 4.2.12 that such sequences are called 'dominating sequences' for H^∞.

(1) $|f(z)| \leq M(1/(1 - |z|))$.

(2) *Every point of* $\mathbb{T}$ *is the nontangential limit of a sequence of zeros of* f.

PROOF. First observe that the polynomial $P_n(z) := 1 - 2z^n$ has n equally spaced zeros on the circle of radius

$$\exp(-(\log 2)/n) \asymp 1 - (\log 2)/n$$

for large n. If now E is any infinite set of positive integers, the function

$$f_E(z) := \prod_{n \in E} (1 - 2z^n)$$

(which converges for $z \in \mathbb{D}$) is analytic in $\mathbb{D}$. A routine exercise in geometry proves statement (2) of the proposition.

To prove statement (1), note that for $z \in \mathbb{D}$,

$$|f_E(z)| \leq \prod_{n \in E} (1 + 2|z|^n) \leq \exp\left(2 \sum_{n \in E} |z|^n \right).$$

With the choice of a sufficiently 'sparse' set E [**52**, p. 87], we can arrange things such that

$$2 \sum_{n \in E} |z|^n \leq \log M(1/(1 - |z|)).$$

$\square$

REMARK 6.6.5. Choosing $M(x) = x^p$ for some $p < 1/2$ gives our assertion concerning the Bergman space.

Returning to our construction, we can appropriately choose non-zero coefficients c_n so that the Borel series

(6.6.6)
$$f = \sum_{n=1}^{\infty} \frac{c_n}{1 - \overline{a_n} z}$$

converges in $\mathcal{D}$ to a non-trivial function. Without loss of generality, assume also that $\{c_n\}$ is absolutely summable and so the above series converges uniformly on compact subsets of $\mathbb{C} \backslash \{1/\overline{a_n}\}^-$. Since $\mathcal{M}(A) \neq \mathcal{D}$, the function f is a non-cyclic vector for B. The claim here is that the coefficients $\{c_n\}$ can be chosen to converge to zero so fast that if $f|\mathbb{D}$ has a pseudocontinuation across any arc of $\mathbb{T}$, then this pseudocontinuation must be none other than $f|\mathbb{D}_e$.[4] However, $f|\mathbb{D}_e$ cannot possibly possess non-tangential limits on any arc of the circle since, by design, the poles $\{1/\overline{a_n}\}$ of f accumulate non-tangentially at every boundary point.

Though there are some other partial results about the cyclic vectors for the Dirichlet space (see [**11**] as well as Corollary 8.5.3 of these notes), a complete description of them is very much an open problem.

QUESTION 6.6.7. What are the cyclic vectors for the backward shift on the Dirichlet space?

---

[4]The details of this, found in [**11**, Thm. 6.2], are quite technical and are beyond the scope of what we want to do here.

For the Hardy space, we know from Theorem 6.3.8 that the sum and product of two non-cyclic vectors is non-cyclic. Is the same result true for the Bergman and Dirichlet spaces? This issue will be addressed in Chapter 8, § 8.3.

6.7. The Darlington synthesis problem

Let $\Phi = (a_{i,j})$ be an $n \times n$ matrix of bounded analytic functions on the disk. Φ is said to be a 'matrix-valued inner function' if the map

$$\begin{pmatrix} f_1 \\ \vdots \\ f_n \end{pmatrix} \to \Phi \begin{pmatrix} f_1 \\ \vdots \\ f_n \end{pmatrix},$$

is an isometry from $H^2 \oplus \cdots \oplus H^2$ to itself. A well-known result [118, p. 97] says that Φ is matrix-valued inner if and only if for almost every θ, the boundary function $\Phi(e^{i\theta})$ is a unitary matrix. In this section, we wish to demonstrate a relationship between pseudocontinuations and the study of these matrix-valued inner functions.

For the sake of simplicity, we will restrict our discussion to the special case of 2×2 matrices. More precisely, let S, A, B, C be bounded analytic functions on $\mathbb{D}$ such that

$$U := \begin{pmatrix} S & B \\ A & C \end{pmatrix}$$

is a 2×2 matrix-valued inner function, that is to say, for almost every θ,

$$U(e^{i\theta})U^*(e^{i\theta}) = \begin{pmatrix} 1 & 0 \\ 0 & 1 \end{pmatrix}.$$

Notice that the $(1,1)$ entry of $U(e^{i\theta})U^*(e^{i\theta})$ is equal to

$$|S(e^{i\theta})|^2 + |B(e^{i\theta})|^2 = 1 \text{ for a.e. } \theta$$

and so $|S| \leq 1$ on $\mathbb{D}$. Also notice that since

$$U(e^{i\theta}) = (U^*(e^{i\theta}))^{-1} \text{ a.e. } \theta,$$

then

$$\begin{pmatrix} S & B \\ A & C \end{pmatrix} = \begin{pmatrix} \overline{C/D} & -\overline{B/D} \\ -\overline{A/D} & \overline{S/D} \end{pmatrix} \text{ a.e. on } \mathbb{T},$$

where $D = SC - BA$. Thus $S = \overline{C/D}$ a.e on $\mathbb{T}$ and so

$$T_S(z) := (\overline{C/D})(1/\bar{z})$$

is a pseudocontinuation of S (across $\mathbb{T}$) of bounded type. Using our previous notation (see Definition 6.3.3), we say that $|S| \leq 1$ and $S \in PCBT$. Pretty much the same argument shows that if $\Phi = (a_{i,j})$ is an $n \times n$ matrix-valued inner function, then $|a_{1,1}| \leq 1$ and $a_{1,1} \in PCBT$.

The more difficult question (known in electrical engineering circles as the 'Darlington synthesis problem') is the converse. Namely, if $|S| \leq 1$ is analytic on $\mathbb{D}$ and belongs to $PCBT$, do there exist $A, B, C \in H^\infty(\mathbb{D})$ such that

$$U = \begin{pmatrix} S & B \\ A & C \end{pmatrix}$$

is a matrix-valued inner function? The answer to this question, due to Arov [15] and Douglas and Helton [49], is yes.

THEOREM 6.7.1. *Let $S \in H^\infty(\mathbb{D})$, $|S| \leq 1$, with $S \in PCBT$. Then there are functions $A, B, C \in H^\infty(\mathbb{D})$ such that*

$$U = \begin{pmatrix} S & B \\ A & C \end{pmatrix}$$

is a matrix-valued inner function.

The proof of this theorem will depend on the following two lemmas.

LEMMA 6.7.2. *A function $f \in L^\infty(\mathbb{T})$ is the boundary function for a meromorphic function on $\mathbb{D}_e$ of bounded type if and only if there is an inner function ϕ on $\mathbb{D}$ such that $\phi \bar{f} \in H^\infty(\mathbb{T})$* [5].

PROOF. Suppose $\phi \bar{f} = h \in H^\infty(\mathbb{T})$. Then $f(e^{i\theta}) = \bar{h}(e^{i\theta})/\bar{\phi}(e^{i\theta})$ almost everywhere and so

$$T_f(z) := \overline{(h/\phi)}(1/\bar{z}), \quad |z| > 1,$$

is of bounded type (on the exterior disk) and has boundary values equal to f.

Conversely, suppose $f(e^{i\theta}) = F_1(e^{i\theta})/F_2(e^{i\theta})$ a.e., where $F_1, F_2 \in H^\infty(\mathbb{D}_e)$. Then, setting

$$f_j(z) := \overline{F_j(1/\bar{z})}, \quad j = 1, 2, \quad z \in \mathbb{D},$$

one can see that $f(e^{i\theta}) = \overline{f_1}(e^{i\theta})/\overline{f_2}(e^{i\theta})$ a.e., where $f_1, f_2 \in H^\infty(\mathbb{T})$. Factor $f_j = O_j I_j$, where O_j is outer and I_j is inner, and observe that

$$\bar{f} = \frac{O_1}{O_2}\frac{I_1}{I_2} \quad \text{a.e. on } \mathbb{T}.$$

Thus

$$I_2 \bar{f} = \frac{O_1}{O_2} I_1 \in L^\infty(\mathbb{T}) \cap N^+(\mathbb{T}),$$

where $N^+(\mathbb{T})$ are the boundary functions of the Smirnov class N^+. It follows from standard theory [**52**, Thm. 2.11] that $I_2 \bar{f} \in H^\infty(\mathbb{T})$. $\qquad\square$

LEMMA 6.7.3. *Suppose that $S \in H^\infty$, $|S| \leq 1$, $S \in PCBT$, and S is not inner. Then there is a bounded outer function A such that $|A(e^{i\theta})|^2 = 1 - |S(e^{i\theta})|^2$ almost everywhere.*

PROOF. Since $S \in PCBT$, then $S = \phi \bar{g}$ on $\mathbb{T}$ for some inner function ϕ and $g \in H^\infty(\mathbb{T})$ (Lemma 6.7.2). Hence

$$|S|^2 = \phi \bar{S} \bar{g} = \bar{\phi} S g = S g / \phi$$

on $\mathbb{T}$ and so

$$1 - |S|^2 = \frac{\phi - Sg}{\phi}.$$

Moreover, $\phi - Sg$ is not the zero function since if it were, then

$$\phi = Sg = (\phi \bar{g})g$$

almost everywhere on $\mathbb{T}$ and so $|g| = 1$ a.e., implying S is inner, a contradiction. Thus

$$\log(1 - |S|^2) = \log|\phi - Sg|$$

---

[5] $H^\infty(\mathbb{T}) = \{f \in L^\infty(\mathbb{T}) : f(e^{i\theta}) = \lim_{r \to 1^-} f(re^{i\theta}) \text{ a.e.}, f \in H^\infty(\mathbb{D})\}$

which, by a theorem of Riesz [**60**, Thm. 4.1] and the fact that $\phi - Sg$ is a non-trivial function from $H^\infty(\mathbb{T})$, is integrable. The function

$$A(z) := \exp\left(\int_0^{2\pi} \frac{e^{i\theta} + z}{e^{i\theta} - z} \frac{1}{2} \log(1 - |S(e^{i\theta})|^2) \frac{d\theta}{2\pi}\right)$$

(see eq.(2.2.5)) is a bounded outer function with $|A|^2 = 1 - |S|^2$ almost everywhere on $\mathbb{T}$ [**60**, pp. 66 - 67]. $\qquad\square$

PROOF OF THEOREM 6.7.1: We follow the proof of Douglas and Helton [**49**, pp. 65 - 66]. To avoid trivial cases, we will assume that S is not an inner function, since if it were, we could set $A = B = 0$ and $C = 1$.

Since $S \in PCBT$, $|S| \le 1$, and not inner, we can apply Lemma 6.7.3 to produce a bounded outer function A such that $|A|^2 = 1 - |S|^2$ almost everywhere on $\mathbb{T}$. Let

$$C_0 := -A^2 \frac{1}{1 - |S|^2} \overline{S}$$

and notice that $1 - |S|^2 = |A|^2$ is not equal to zero on any set of positive measure (since A is an outer function) and so there is no ambiguity in the (almost everywhere) definition of C_0. By our choice of A, the function C_0 is bounded, but unfortunately may not belong to $H^\infty(\mathbb{T})$. To rectify this, notice that $\overline{S}$ is the boundary function of $\overline{S}(1/\overline{z})$, $z \in \mathbb{D}_e$. Thus we may apply Lemma 6.7.2 to produce an inner function ϕ_1 such that $\phi_1 \overline{S} \in H^\infty(\mathbb{T})$. If T_S is the pseudocontinuation (of bounded type) of S, then

$$1 - T_S(z)\overline{S}(1/\overline{z})$$

is a function of bounded type on $\mathbb{D}_e$ with boundary function equal to $1 - |S(e^{i\theta})|^2$. Again apply Lemma 6.7.2 to produce an inner ϕ_2 such that $\phi_2(1 - |S|^2) \in H^\infty(\mathbb{T})$. Factor $\phi_2(1 - |S|^2) = IO$, where I is inner and O is outer and notice that

$$C := I\phi_1 C_0 = -A^2 \frac{\phi_2}{O}\overline{S}\phi_1 \in L^\infty(\mathbb{T}) \cap N^+(\mathbb{T}) = H^\infty(\mathbb{T}).$$

Setting $B := I\phi_1 A$, we leave it to the reader to show that

$$U = \begin{pmatrix} S & B \\ A & C \end{pmatrix}$$

has unitary boundary values. $\qquad\square$

As mentioned earlier, the above unitary embedding problem, and various generalizations [**15**, **49**], fall under the rubric of 'Darlington synthesis' and have important connections to electrical engineering [**44**, **102**].

6.8. Linear differential equations of infinite order

One can regard the study of cyclic vectors for forward and backward shifts as a chapter in the study of infinite systems of linear equations, a far from complete theory initiated by H. Poincaré, H. von Koch [**142**], and F. Riesz [**115**]. For example, to say that a sequence $a = \{a_0, a_1, \ldots\}$ in ℓ^2 is a cyclic vector for the forward shift

$f \to zf$ on H^2 is equivalent to saying the only ℓ^2 vector $b = \{b_0, b_1, \ldots\}$ satisfying the linear system (written in the usual matrix - vector notation)

$$(6.8.1) \qquad \begin{pmatrix} a_0 & a_1 & a_2 & \cdots \\ 0 & a_0 & a_1 & \cdots \\ 0 & 0 & a_0 & \cdots \\ \vdots & \cdots & \cdots & \vdots \end{pmatrix} \begin{pmatrix} b_0 \\ b_1 \\ \vdots \\ \vdots \end{pmatrix} = \begin{pmatrix} 0 \\ 0 \\ \vdots \\ \vdots \end{pmatrix}$$

is the zero vector. Likewise, the vector a is a cyclic vector for the backward shift $f \to (f - f(0))/z$ on H^2 if and only if $b = 0$ is the only ℓ^2 solution of the corresponding system where the matrix on the left hand side of eq.(6.8.1) is replaced by

$$(6.8.2) \qquad \begin{pmatrix} a_0 & a_1 & a_2 & \cdots \\ a_1 & a_2 & a_3 & \cdots \\ a_2 & a_3 & a_4 & \cdots \\ \vdots & \cdots & \cdots & \vdots \end{pmatrix}.$$

The matrix in eq.(6.8.1) is a 'Toeplitz matrix', since the entries on each line parallel to the main diagonal are equal, while the one in eq.(6.8.2) is a 'Hankel matrix', since the entries on each line perpendicular to the main diagonal are equal. A great part of the special structure inherent in each of these classes of cyclic vectors is due to the Toeplitz (respectively Hankel) structure of these matrices.

Now, infinite systems of linear equations whose matrices are of the Toeplitz or Hankel form occur in many areas of analysis. Without going into details here, which would lead us too far afield, let us give one important example, coming from the subject of linear differential equations of infinite order with constant coefficients as explored by Ritt [**116**], Pólya [**107**], Valiron [**141**] and Cramér [**39**]. This subject has a long history - and an immense bibliography - of which we only can refer here to a small part. Beside the papers mentioned, the reader could consult the survey articles of Carmichael [**33, 34**], Leont'ev [**93**], and Muggli [**101**].

DEFINITION 6.8.3. Suppose f is an analytic function on a disk

$$G = \{|z| < R\}, \ 0 < R \leq \infty,$$

with Taylor expansion about $z = 0$

$$f(z) = a_0 + a_1 z + a_2 z^2 + \cdots .$$

If ϕ is any formal power series,

$$\phi(w) = c_0 + c_1 w + c_2 w^2 + \cdots ,$$

and D denotes the differential operator

$$D = \frac{d}{dz},$$

we will say $\phi(D)$ 'operates on f', if the result of formally applying the (in general, infinite order) differential operator

$$\phi(D) = c_0 I + c_1 D + c_2 D^2 + \cdots$$

to f, term by term, gives a series absolutely convergent in G. That is, we require

$$(6.8.4) \qquad \sum_{0 \leq j \leq k < \infty} |c_j||a_k|k(k-1)(k-2)\cdots(k-j+1)|z|^{k-j} < \infty \ \text{ for all } z \in G.$$

When this holds,

$$(\phi(D)f)(z) = \sum_{j=0}^{\infty} c_j f^{(j)}(z)$$

is again a function analytic in G and all computations based on reordering of series will be valid.

Usually one is interested in two 'classes' of power series, say $\{f\}$ and $\{\phi\}$, such that all those in $\{f\}$ are analytic in some disk G, and $\phi(D)$ operates on f for each $\phi \in \{\phi\}$ and $f \in \{f\}$. The most studied case is where $\{f\}$ is the set E of all entire functions and $\{\phi\}$ is the set X of all entire functions of exponential type (see Chapter 2). To motivate our later discussion, we review this set-up briefly. We leave it to the reader to check [**39, 101**] for the appropriate details.

PROPOSITION 6.8.5. *If $\phi \in X$ and $f \in E$, then $\phi(D)f \in E$. Moreover, the mapping $f \to \phi(D)f$ is continuous on E.*

It is also true that every operator $f(D)$ with f in E operates on all functions in X. Notice the trend: X is a much smaller class than E, so the set of (infinite order) differential operators which operate on X is a much larger class. In the case we shall study in detail below, the functions being differentiated will be a subset of X (provided with a Hilbert space norm) and the symbols of the operators acting on it will comprise H^∞. But first, let us examine the above E, X scenario. Consider the following question:

QUESTION 6.8.6. Given a function f in E, does there exist a non-trivial $\phi \in X$, *i.e.*, ϕ not identically zero, such that

$$\phi(D)f = 0?$$

We state without proof the following.

PROPOSITION 6.8.7. *For a non-trivial $f \in E$, the following are equivalent.*
(1) *The only $\phi \in X$ such that $\phi(D)f = 0$ is $\phi = 0$.*
(2) *The translates*

$$\{f(z+p) : p \in \mathbb{C}\}$$

span E in its natural topology of uniform convergence on compacta.
(3) *The sequence of derivatives $f, f', f'', \dots$ spans E.*

DEFINITION 6.8.8. An entire function for which the translates

$$\{f(z+p) : p \in \mathbb{C}\}$$

fail to span E in its natural topology of uniform convergence on compacta is said to be 'mean periodic'.

Before proceeding, let us say a few words about the general theory of mean periodic functions, which is very closely related to cyclic vectors of operators. Delsarte [**41**] introduced this notion in the context of the space $C(\mathbb{R})$ of continuous functions on $\mathbb{R}$, endowed with the topology of uniform convergence on compacta. A function $f \in C(\mathbb{R})$ is said to be 'mean periodic' if its translates

$$\{f_y(x) := f(x - y) : y \in \mathbb{R}\}$$

fail to span $C(\mathbb{R})$ (roughly speaking, if f fails to be a cyclic vector with respect to the action of the translation group). This is easily seen to be equivalent to the existence of a nontrivial compactly supported complex measure μ on $\mathbb{R}$ obeying the convolution equation

$$(f * \mu)(y) = \int f(x - y) d\mu(x) = 0 \quad \text{for all } y.$$

This convolution equation generally has exponential solutions, namely, $f(x) = e^{wx}$, where w is any complex zero of the Laplace transform

$$M(w) := \int e^{wx} d\mu(x)$$

of μ. Moreover, in the case where M has multiple zeros, there are corresponding solutions where the exponentials are multiplied by appropriate powers of x. These products have been called 'monomial-exponentials'. From this context, several problems naturally arise:

(1) Are all solutions f to $f * \mu = 0$ spanned ('synthesizable') by linear combinations of the aforementioned exponential and monomial-exponential solutions? The answer turns out to be yes - although highly non-trivial to prove. Without attempting to give an exhaustive review of the massive literature dealing with this and other questions, we refer the reader to J.-P. Kahane's thesis [82] (see also [83]).

(2) For a given f which is mean periodic, is there a canonical family of exponentials (and monomial-exponentials) that spans f? What is meant by this vague and imprecise question is the following: If f is mean periodic, then there are (many) nontrivial compactly supported measures μ with $f * \mu = 0$, and by (1), each of these gives rise to a family of exponentials spanning f. When we say this, we henceforth also include the monomial-exponentials, if applicable. But, for any given μ with $f * \mu = 0$, we may not need *all* of the exponentials corresponding to the zeros of $M = 0$. Is there perhaps a *minimal* choice of μ such that no proper subset of the zeros of M give a spanning set of exponentials?

(3) Can we obtain a 'canonical' set of exponentials for spanning a given mean periodic f, by choosing those exponentials that in turn are spanned by the translates of f? This is motivated by Beurling's posing of the analogous 'spectral synthesis' question in $L^\infty(\mathbb{R})$ [20, 47].

(4) If there is (in some suitable sense) a 'canonical' (and hence minimal) set of exponentials associated to f, and spanning f, can we go further and assign 'Fourier coefficients' to these exponentials so the resulting series represents f, in the sense say of uniform convergence on compacta? For example: if f has period 1, it certainly is mean periodic. There is a natural choice of measure μ here, which is atomic, placing masses 1 and -1 at $x = 1$ and $x = 0$ and the associated exponentials are

$$\{e^{2\pi inx} : n \in \mathbb{Z}\}.$$

These indeed span all the functions of period 1. But for a given f, not all of them may be needed (*i.e.*, in case f has one or more vanishing Fourier coefficients, which can be shown equivalent to f's translates not spanning the corresponding exponentials).

We have reviewed this picture because it is a good 'model' for a very general situation that arises in analysis in various connections (including cyclic vectors for shift operators). This was recognized by L. Schwartz, whose programmatic paper [**122**] placed on the order of the day, so to speak, questions of 'spectral synthesis' in a wider variety of contexts. Delsarte had introduced the notion of mean periodic functions in $C(\mathbb{R})$ and even $C(\mathbb{R}^n)$. But, in $C(\mathbb{R}^n)$ with $n > 1$ say, every f which is harmonic is trivially mean periodic. What is really of interest for a harmonic function is not whether its translates span $C(\mathbb{R}^n)$ (obviously they don't) but whether they span all the harmonic functions contained in $C(\mathbb{R}^n)$. Those that do not are, in L. Schwartz's terminology, the 'mean periodic harmonic functions'. In like manner, an entire f on $\mathbb{C}$ is mean periodic if its translates do not span the entire functions. In all these contexts, the above thematic questions (1) - (4), and some others, impose themselves and their answers in some cases, especially in multivariable situations, are still being sought. Although the cyclic vector problems which interest us in the present study are only a small corner of the large picture, it seems worthwhile to see it against the backdrop of more general questions.

After this digression, let us focus again on mean periodic entire functions. In a weak form, spectral synthesis can be seen as a property pertaining to a given $\phi \in X$. Namely, $\phi(D)f = 0$ has, in general, 'exponential' solutions of the type

$$f(z) = e^{wz}.$$

Indeed, this is a solution if and only if $\phi(w) = 0$, and, there is an obvious generalization for zeros of ϕ of higher multiplicity:

$$f(z) = p(z)e^{wz}$$

is a solution, for every polynomial p of degree at most $m - 1$, if and only if w is an m-fold zero of ϕ. This collection, corresponding to the zeros of ϕ, is called the family of 'exponential solutions'.

DEFINITION 6.8.9. 'Weak spectral synthesis' holds for ϕ, if and only if every solution $f \in E$ to $\phi(D)f = 0$ is in the closed linear span of the exponential solutions for ϕ.

For a given mean periodic $f \in E$, its family of associated exponentials is the set of functions of the form e^{wz} which are spanned by the translates (equivalently, by the derivatives) of f. It is easy to see that if $p(z)e^{wz}$ is in this class, so is $z^m e^{wz}$ for every non-negative integer m not exceeding the degree of p.

DEFINITION 6.8.10. We say that 'strong spectral synthesis' holds for f, if f is in the span of its family of associated exponentials.

THEOREM 6.8.11. *For the spaces X and E,*

(1) *(Ritt, Valiron, and others) Weak spectral synthesis holds for every $\phi \in X$.*
(2) *(L. Schwartz) Strong spectral synthesis holds for every $f \in E$.*

REMARK 6.8.12. More refined versions were given later by A. Gel'fond [61] and H. S. Shapiro [126]. Although we won't pursue them here, multivariable generalizations [98] are much deeper and have attracted great interest.

Coming back to the notion of mean periodicity, the entire function f is mean periodic if and only if $\phi(D)f = 0$ for a non-trivial $\phi \in X$. Let us explore this a bit further. If $\phi(D)f = 0$, then, denoting by $\{b_n\}$ the Taylor coefficients of ϕ,

$$\sum_{n=0}^{\infty} b_n D^n f = 0,$$

and it is easy to see that this can be differentiated termwise to yield

$$\sum_{n=0}^{\infty} b_n D^{n+1} f = 0,$$

and differentiating repeatedly yields

$$\sum_{n=0}^{\infty} b_n D^{n+m} f = 0, \quad \text{for all } m = 0, 1, 2, \ldots$$

Thus, if we denote

$$A_m := m! a_m = (D^m f)(0),$$

the last equation (evaluated at $z = 0$) can be written as

(6.8.13) $$\sum_{n=0}^{\infty} b_n A_{n+m} = 0, \quad m = 0, 1, 2, \ldots,$$

or equivalently in matrix form as

$$\begin{pmatrix} A_0 & A_1 & A_2 & \cdots \\ A_1 & A_2 & A_3 & \cdots \\ A_2 & A_3 & A_4 & \cdots \\ \vdots & \cdots & \cdots & \vdots \end{pmatrix} \begin{pmatrix} b_0 \\ b_1 \\ \vdots \\ \vdots \end{pmatrix} = \begin{pmatrix} 0 \\ 0 \\ \vdots \\ \vdots \end{pmatrix}.$$

Observe that formally, this is a system like the one encountered in studying cyclic vectors of the backward shift: as a system of equations for the 'unknowns' $\{b_n\}$, its matrix is a Hankel matrix involving the A_n's (compare with eq.(6.8.2) above). Moreover, it is not hard to show the steps leading to eq.(6.8.13) are reversible, and if eq.(6.8.13) has a non-trivial solution in which the sequence $\{b_n\}$ are the Taylor coefficients of an entire function of exponential type, that is

$$n! b_n = O(C^n)$$

for some constant C, then ϕ having the Taylor coefficients $\{b_n\}$ satisfies

$$\phi(D)f = 0.$$

Of course, we can, in examining the system of equations in eq.(6.8.13), also 'put the shoe on the other foot' and look upon it as if A_n's were the unknowns and the matrix of the system is a Toeplitz matrix whose entries are the b_n's.

Also, just as the backward shift operator plays a role dual to that of the forward shift operator, the derivative D plays a role dual to that of the integration operator

$$(Jf)(z) = \int_0^z f(t)\,dt$$

which, like D, is a continuous linear map from E to E. Consider, for example, the question: for a given $f \in E$, does the set

$$\{J^m f : m = 0, 1, 2, \ldots\}$$

span E? The necessary and sufficient condition for this to happen is that, for every

$$\phi = \sum_{n=0}^{\infty} b_n w^n \in X,$$

the relations

$$[\phi, J^m f] = 0 \quad m = 0, 1, 2, \ldots$$

imply $\phi = 0$. Recall from eq.(2.2.8), the duality bracket $[\cdot, \cdot]$ here is defined by

$$[\phi, f] := \sum_{n=0}^{\infty} n! b_n a_n = (\phi(D)f)(0).$$

Since

$$J^m f = \sum_{n=0}^{\infty} \frac{a_n z^{n+m}}{(n+1)(n+2) \cdots (n+m)}, \quad m = 0, 1, 2, \ldots,$$

this condition becomes

(6.8.14)
$$\sum_{n=0}^{\infty} b_{n+m} n! a_n = 0, \quad m = 0, 1, 2, \ldots.$$

This is, in terms of the numbers $A_n := f^{(n)}(0) = n! a_n$, the matrix equation

$$\begin{pmatrix} A_0 & A_1 & A_2 & \cdots \\ 0 & A_0 & A_1 & \cdots \\ 0 & 0 & A_0 & \cdots \\ \vdots & \cdots & \cdots & \vdots \end{pmatrix} \begin{pmatrix} b_0 \\ b_1 \\ \vdots \\ \vdots \end{pmatrix} = \begin{pmatrix} 0 \\ 0 \\ \vdots \\ \vdots \end{pmatrix},$$

which is the analogous system as before, see eq.(6.8.13), but with the Hankel matrix replaced by a Toeplitz matrix.

In order to bring these ideas into closer contact with those discussed in this book (continuation properties of analytic functions), we shall now examine another pair of spaces of analytic functions in place of E and X which are related to the Hardy space H^2, so that various known results about cyclic vectors in H^2 translate into equivalent, but formally quite different results concerning infinite order differential equations. We limit our study to this one example.

Let us define a Hilbert space V of entire functions

$$F(z) = A_0 + A_1 z + \cdots$$

such that $\{n! A_n\} \in \ell^2$. The norm is given by

$$\|F\| := \sqrt{\sum_{n=0}^{\infty} (n!)^2 |A_n|^2}.$$

It is easy to check that V is a (proper) subset of the entire functions of exponential type at most one. We shall give an integral formula for $\|F\|$ later. It is straightforward to check that the mapping $U : H^2 \to V$ defined by

$$f = \sum_{n=0}^{\infty} a_n z^n \to Uf := \sum_{n=0}^{\infty} \frac{a_n}{n!} z^n$$

is a unitary operator and moreover, a computation with Taylor series shows that

$$UB = DU,$$

where $B = S^*$ is the usual backward shift operator on H^2.

Suppose now that f is a non-cyclic vector for the backward shift on H^2. Then, in view of the isomorphism between H^2 and V, the vector $F := Uf$ is 'non-cyclic' in the sense that the family

$$F, F', F'', \ldots$$

does not span V (in its norm topology). Conversely, if for a given $F \in V$, these derivatives fail to span V, then $f := U^{-1}F$ is non-cyclic for the backward shift. It is well-known that if $\phi \in H^{\infty}$, then the operator $\phi(B)$ is well defined and equal to the co-analytic Toeplitz operator $T_{\bar{\phi}}$ on H^2. Here if $g \in L^{\infty}(\mathbb{T})$, the operator

$$T_g f := P(gf),$$

where P is the 'Riesz projection'

$$(6.8.15) \qquad P\left(\sum_{n=-\infty}^{\infty} \hat{f}(n) e^{in\theta} \right) = \sum_{n=0}^{\infty} \hat{f}(n) e^{in\theta}$$

of $L^2(\mathbb{T})$ onto $H^2(\mathbb{T})$, is the 'Toeplitz operator' on H^2 with symbol g. Moreover, one can show that $\|T_g\| \le \|g\|_{\infty}$. From the unitary equivalence of the operators B and D, one notes that $\phi(D)$ is a bounded operator on V with norm not exceeding the supremum norm of ϕ on $\mathbb{D}$. It is also an elementary fact that $f \in H^2$ is non-cyclic for B if and only if there is a non-trivial $\phi \in H^{\infty}$ such that $f \in \ker T_{\bar{\phi}}$. In fact, every B-invariant subspace $\mathcal{M}$ of H^2 takes the form $\mathcal{M} = \ker T_{\bar{\phi}}$ for some inner ϕ. Putting these facts together, we obtain the following summary result.

THEOREM 6.8.16.   (1) *For $F \in V$,*

$$\bigvee \{F, F', F'', \ldots\} \ne V$$

if and only if F belongs to the kernel of a non-trivial infinite order differential operator $\phi(D)$ with $\phi \in H^{\infty}$.
(2) *A function $F \in V$ belongs to $\ker \phi(D)$, for any non-trivial $\phi \in H^{\infty}$, if and only if $U^{-1}F$ is non-cyclic for B.*
(3) *Every D-invariant subspace $\mathcal{F}$ of V takes the form $\mathcal{F} = \ker \phi(D)$ for some inner ϕ.*

As a simple application of these ideas, we have the following example.

EXAMPLE 6.8.17. Let

$$F(z) = \frac{e^z - 1}{z} = \sum_{n=0}^{\infty} \frac{z^n}{(n+1)!}.$$

Observe that

$$U^{-1}F = \sum_{n=0}^{\infty} \frac{z^n}{n+1} = \frac{1}{z} \log(\frac{1}{1-z})$$

is a cyclic vector for B on H^2 due to the isolated winding point at $z = 1$ (Remark 6.3.5). Thus the successive derivatives of F span V. Moreover, F is not in the kernel of $\phi(D)$ for any non-trivial $\phi \in H^{\infty}$.

We can state analogous results where f being cyclic for the forward shift on H^2 is equivalent to the successive antiderivatives

$$J^n F, \quad n = 0, 1, 2, \ldots, \quad F = U^{-1}f$$

spanning V. By Beurling's theorem (Theorem 6.3.2), this happens for any outer f, say $f = (1 + z)^2$, whence $F = 1 + 2z + z^2/2$. Thus, the collection

$$F = 1 + 2z + \frac{z^2}{2}, JF = z + z^2 + \frac{z^3}{6}, \ldots$$

spans V, whereas for say $f = 1 + 2z$, here $F = 1 + 2z$, the sequence

$$F = 1 + 2z, JF = z + z^2, \ldots$$

fails to span V since f is not outer.

The analogy with spectral synthesis can also be pursued here. For an exponential function $F(z) = e^{az}$ belonging to V, where necessarily $a \in \mathbb{D}$,

$$U^{-1}F = \frac{1}{1 - az},$$

and so the 'reproducing kernels' of the Hardy space correspond to exponentials in V. As we know, spectral synthesis (recall Definition 6.5.3) for the backward shift operator B with root vectors

$$k_{\lambda}^{(n)}(z) := \frac{n!z^n}{(1 - \bar{\lambda}z)^{n+1}},$$

fails miserably in H^2. That is to say, there are B-invariant subspaces $\mathcal{M}$ which cannot be written as

$$\mathcal{M} = \bigvee \left\{ k_{\lambda}^{(n)} : k_{\lambda}^{(n)} \in \mathcal{M} \right\}.$$

Recall that such a 'synthesizable' $\mathcal{M}$ (spanned by its root vectors) must be of the form $(bH^2)^{\perp}$, where b is a Blaschke product. Likewise, spectral synthesis fails in V, i.e., solutions to $\phi(D)F = 0$ are not, in general, approximable by linear combinations of exponential solutions.

We end this section with a comment about an equivalent definition of the space V. We claim that the norm on V can be expressed as

(6.8.18) $$\|F\|^2 \asymp \int_{\mathbb{C}} |F|^2 W \, dA,$$

where W is a certain weight function and dA is two-dimensional area measure. To see this, let $W(r)$ be a positive function on $(0, \infty)$, and consider, for each entire F, the expression

$$N(F, W) := \int_{\mathbb{C}} |F(z)|^2 W(|z|) \, dA.$$

In terms of the Taylor coefficients $\{A_n\}$ of F, the quantity $N(F, W)$ is equal to

$$2\pi \sum_{n=0}^{\infty} p_n |A_n|^2,$$

where

(6.8.19)
$$p_n = \int_0^{\infty} r^{2n+1} W(r) dr.$$

If we could choose the weight W so that this integral were a constant multiple of $(n!)^2$, we would have an integral representation for $\|F\|^2$ of the form in eq.(6.8.18). From the theory of the Stieltjes moment problem [**3**, p. 76], one knows there is such a solution W. But, without entering into these subtleties, let us note that the choice

$$W(r) = e^{-tr}$$

yields

$$p_n = (2n+1)! t^{-2(n+1)}$$

and so

(6.8.20)
$$\frac{p_n}{(n!)^2} = (2n+1) \left[\frac{(2n)!}{(n!)^2} \right] t^{-2(n+1)}.$$

Now, the bracketed term is asymptotically a constant times $4^n/\sqrt{n}$ (by Stirling's formula [**109**, p. 44]). Hence for $W(r) = e^{-2r}$, the right side of eq.(6.8.20) is asymptotically $C\sqrt{n}$, and so there is a constant K such that

$$\|F\|^2 \leq K \int_{\mathbb{C}} |F(z)|^2 e^{-2|z|} dA$$

for all $F \in V$. So, all entire functions square integrable with respect to the weight $e^{-2|z|}$ are in V. However, this space is strictly smaller than V. By similar reasoning, V is strictly contained in the space of entire functions square integrable with respect to the weight $e^{-t|z|}$ for every $t < 2$.

Thus, to sum up: In a certain Hilbert space V of entire functions 'close' to those square integrable with respect to the weight $e^{-2|z|}$, a series of classical questions concerning infinite order ordinary differential equations, mean periodicity, spectral synthesis, etc., are equivalent to questions about the backward shift operator on H^2 and hence involve the concept of pseudocontinuation. Presumably this analogy can be pushed further: on the one hand, replacing H^2 by other spaces, and on the other, (even in H^2) interpreting the meaning of the $C_{f,g}$ transform (see eq.(6.3.6)) and other objects associated to B-invariant subspaces of H^2 in the context of the associated D-invariant subspaces of V.

6.9. Gap theorems

Gap series are Taylor series of the form

(6.9.1)
$$\sum_{n=0}^{\infty} a_n z^{\lambda_n}, \quad \lambda_0 < \lambda_1 < \lambda_2 < \cdots,$$

where, in one sense or another, the λ_n's are a 'scarce' subset of the integers. They have enjoyed a long history in complex analysis and play an important role in understanding analytic continuation of Taylor series across their circles of convergence (which we assume here to be the unit circle) [**45, 137**] .

In 1872, Weierstrass [144] gave his famous example of a Fourier gap series

$$\sum_{n=1}^{\infty} a^n \cos(\lambda^n \theta),$$

(where $\lambda \geq 3$ is an odd integer, $0 < a < 1$, with $a\lambda > 1 + 3\pi/2$) which is a continuous but nowhere differentiable function of the real variable θ. As a side note, the somewhat mysterious condition $a\lambda > 1 + 3\pi/2$ can be relaxed to $a\lambda \geq 1$ by a result of Hardy [67]. Noticing that the above Weierstrass series is the real part of the Taylor series

$$\sum_{n=1}^{\infty} a^n z^{\lambda^n}$$

evaluated on the circle, one has an example of a Taylor series for which the unit circle is a natural boundary. Recall that $\mathbb{T}$ is a 'natural boundary' for an $f \in \mathfrak{H}(\mathbb{D})$ if f does not have an analytic continuation across any point of $\mathbb{T}$.

In 1890, Fredholm [58, 59] suggested that the failure of differentiability on the boundary is not the only reason why a Taylor series can have the unit circle as a natural boundary. Indeed, there are nowhere continuable series having radius of convergence equal to one for which the derivatives of all orders extend continuously to the closed unit disk. Such a series is

$$\sum_{n=0}^{\infty} a^n z^{n^2}$$

for some $|a| < 1$. Fredholm's argument, based on a brilliant and original use of the heat equation, contained an error, which was pointed out and corrected in [88] .

A large class of non-continuable Taylor series was discovered in 1892 by Hadamard [65].

THEOREM 6.9.2 (Hadamard). *A Taylor series of the form in eq.(6.9.1), and with radius of convergence equal to one, and in addition satisfying*

(6.9.3) $$\frac{\lambda_{n+1}}{\lambda_n} \geq q > 1, \ \ n = 0, 1, 2, \ldots,$$

has the unit circle as a natural boundary.

In 1898, Fabry [55] was able to improve Hadamard's gap theorem and thus further expand the class of non-continuable Taylor series.

THEOREM 6.9.4 (Fabry). *A Taylor series of the form in eq.(6.9.1), and with radius of convergence equal to one, which in addition satisfies*

(6.9.5) $$\lim_{n\to\infty} \frac{n}{\lambda_n} = 0,$$

has the unit circle as a natural boundary.

Fabry's gap theorem is about as far as one can go along these lines by virtue of the following 1942 theorem of Pólya [108] (see also Erdös [54]).

THEOREM 6.9.6 (Pólya). *Suppose, for fixed integers $\lambda_0 < \lambda_1 < \lambda_2 < \cdots$, every series with radius of convergence one of the form in eq.(6.9.1) has the unit circle as a natural boundary. Then*

$$\lim_{n \to \infty} \frac{n}{\lambda_n} = 0.$$

We refer the reader to a nice paper of J.-P. Kahane [**84**] for further information about Fourier and Taylor gap series (see also the previously mentioned books of Dienes [**45**] and Titchmarsh [**137**]).

As we have just seen, gap series, with suitable hypotheses, do not have analytic continuations across any arc of the unit circle. As we shall see below, gap series often give us examples of Taylor series which do not have *pseudocontinuations* across any portion of the unit circle. To get started, we recall the precise definition of pseudocontinuation (see Definition 6.2.1): Let I be a non-empty open sub-arc of the unit circle and $f \in \mathfrak{H}(\mathbb{D})$. Then, f has a 'pseudocontinuation' across I if there is an open subset U of $\mathbb{C}$ with $I \subset U$ and a $g \in \mathfrak{M}(U \cap \mathbb{D}_e)$ such that at almost every point of I, f and g possess non-tangential boundary values which are equal.

Notice that we can easily create examples of functions which do not have pseudocontinuations across the whole unit circle. For example, any function with an isolated winding point on the unit circle (*e.g.*, $f(z) = \sqrt{1-z}$) does not have a pseudocontinuation across $\mathbb{T}$ (Example 6.2.3). However, this f still does have pseudocontinuation (in fact an analytic continuation) across *certain* arcs of the unit circle (the ones which avoid the winding point $z = 1$). The point here is to create a function which does not have a pseudocontinuation across *any* arc of the unit circle. As to be expected, our examples come from gap series. We begin with probably the first example of such a series.

THEOREM 6.9.7 (H. S. Shapiro [**131**]). *If*

$$f(z) = \sum_{n=0}^{\infty} 2^{-n} z^{2^n},$$

(*note that f is continuous on $\mathbb{D}^-$ and has radius of convergence equal to one*) *then f does not have a pseudocontinuation across any arc of $\mathbb{T}$.*

To merely show that the H^2 function f in the above theorem does not belong to $PCBT$, that is to say, f is a cyclic vector for the backward shift on H^2 (Theorem 6.3.4), is relatively easy. Indeed, if $f \in PCBT$, there is a $T_f \in \mathfrak{N}(\mathbb{D}_e)$ which is a pseudocontinuation of f across $\mathbb{T}$. Let λ_k denote a primitive 2^k-th root of unity. Then

(6.9.8) $f(z) - f(\lambda_k z) = p_k(z), \quad z \in \mathbb{D},$

for some polynomial p_k. Since the above equation also holds a.e. on $\mathbb{T}$, then

$$T_f(z) - T_f(\lambda_k z) = p_k(z), \quad z \in \mathbb{D}_e.$$

One can check that for a suitable choice of k, the degree of the polynomial p_k can be made as large as desired, contradicting the assumption that T_f and hence $T_f(z) - T_f(\lambda_k z)$ has at worst a pole at infinity. We also mention two general theorems along these lines. The first, due to Douglas-Shapiro-Shields [**51**] says that if

$$f = \sum_{k=1}^{\infty} a_k z^{n_k}$$

belongs to H^2 and

(6.9.9) $$\frac{n_{k+1}}{n_k} \geq d > 1,$$

then $f \notin PCBT$. The second, due to Abakumov [**2**, Thm. 3.3], says the same is true if the sequence of integers $\{n_k\}$ can be written as a finite union of Hadamard gap sequences (*i.e.*, those satisfying eq.(6.9.9)).

The real strength of Theorem 6.9.7 is that f does not have a pseudocontinuation across *any* arc of $\mathbb{T}$ to *any* contiguous domain. The proof will depend on the following lemma.

LEMMA 6.9.10. *Let* $|a| = 1$, $b > 0$ *and let* $U = \{|z - a| < b\} \cap \mathbb{D}_e$. *Suppose that* $n \in \mathbb{N}$ *and so large so that* $(\overline{\lambda}U) \cap U \neq \emptyset$, *where* $\lambda = \exp(2\pi i/n)$. *Let* G *be a meromorphic function on* U *which satisfies, for some polynomial* P, *the functional equation*

(6.9.11) $$G(\lambda z) - G(z) = P(z), \quad z \in (\overline{\lambda}U) \cap U.$$

Then, there is a number $s > 1$ *so that* G *is indefinitely analytically continuable, except possibly for poles, in the annulus* $\{1 < |z| < s\}$. *The extended function has a pole at a point* w *if and only if* $w = \lambda^r w_1$, *where* w_1 *is a pole of* G *in* U *and* $r \in \mathbb{N} \cup \{0\}$. *Furthermore, the necessary and sufficient condition that the extended function be single-valued is that the coefficient of* z^m *in* P *vanishes for all* $m \equiv 0$ *modulo* n.

PROOF. For $r \in \mathbb{N}$, let $U_r = \lambda^{-r}U$. For $z \in U_r$, let

$$G_r(z) := G(\lambda^r z) - \sum_{j=0}^{r-1} P(\lambda^j z).$$

We claim that for every $r \in \mathbb{N}$, G_r is a direct analytic (except for poles) continuation of G_{r-1}. Indeed, when $z \in U_{r-1} \cap U_r$ (which, by hypothesis, is non-empty) we have, from eq.(6.9.11),

$$G(\lambda^r z) - G(\lambda^{r-1} z) = P(\lambda^{r-1} z).$$

Hence, for $z \in U_{r-1} \cap U_r$,

$$G_r(z) - G_{r-1}(z) = G(\lambda^r z) - P(\lambda^{r-1} z) - G(\lambda^{r-1} z) = 0$$

showing that G_r is a direct continuation of G_{r-1} (and hence, ultimately, a continuation of $G_0 = G$). Everything follows once we observe that the singe-valuedness of the continued function is equivalent to the vanishing of $G_n - G$, *i.e.*, the condition that the polynomial

$$\sum_{j=0}^{n-1} P(\lambda^j z)$$

vanishes identically, and this is equivalent to the vanishing of the coefficient in P of z^m for all $m \equiv 0$ modulo n. $\square$

PROOF OF THEOREM 6.9.7. Suppose that

$$f = \sum_{n=0}^{\infty} 2^{-n} z^{2^n}$$

has a pseudocontinuation across some arc of the unit circle. Then there is an $|a| = 1$ and $b > 0$ so that if $U = \{|z - a| < b\} \cap \mathbb{D}_e$ (as in Lemma 6.9.10), then f has an pseudocontinuation (across $U^- \cap \mathbb{T}$) to a function $T_f \in \mathfrak{M}(U)$. Let $n = 2^m$ and $\lambda = \exp(2\pi i/n)$, where m is taken to be large enough so that $(\overline{\lambda} U) \cap U \neq \emptyset$. Then

$$f(\lambda z) - f(z) := P_n(z)$$

is a polynomial of degree less than n. Since $T_f(\lambda z) \in \mathfrak{M}(\overline{\lambda} U)$ is the pseudocontinuation of $f(\lambda z)$ across $\overline{\lambda} U^- \cap \mathbb{T}$, then (by basic properties of pseudocontinuations)

$$T_f(\lambda z) - T_f(z) = P_n(z) \quad z \in (\overline{\lambda} U) \cap U.$$

Applying Lemma 6.9.10 with $G = T_f$, we see that T_f is extendible to a single-valued meromorphic function in an annulus $R = \{1 < |z| < s\}$. Moreover, since we may take for n any sufficiently large power of 2, T_f can have no poles in $R \cap U$, for if T_f had a pole at some point $w_1 \in R \cap U$, the extended function would have poles at a set of points everywhere dense on the circle $\{|z| = |w_1|\}$, which is a contradiction. Therefore T_f is analytic on R, and by Laurent's theorem has a representation $T_f = H + h$, where h is analytic on $\{|z| < s\}$ and H is analytic on $\{|z| > 1\}$ with $H(\infty) = 0$. Now, for every n, we have the functional equation

$$H(\lambda z) + h(\lambda z) = H(z) + h(z) + P_n(z), \quad z \in R.$$

Hence for $z \in R$, we have

$$H(\lambda z) - H(z) = -h(\lambda z) + h(z) + P_n(z)$$

which shows that the function $J(z) := H(\lambda z) - H(z)$ is analytically extendible to $\{|z| < s\}$, and so, since it is analytic in the extended complex plane and vanishes at infinity, $J \equiv 0$. Hence $H(\lambda z) = H(z)$ for all $\lambda = \lambda_n = \exp(2\pi i/n)$, where n is any large power of 2, and so $H \equiv 0$. Therefore $T_f = h$ is analytic on $\{|z| < s\}$. But this implies that f has an analytic continuation to $\{|z| < s\}$, contradicting the fact that f has radius of convergence equal to one. $\square$

REMARK 6.9.12. Shapiro's example extends beyond pseudocontinuations and shows that the gap series f (as in his example) does not admit various other types of 'generalized analytic continuations' such as some of those already covered in these notes, e.g., those of Borel-Walsh and Gončar. One only needs to check that if the functional equation

$$f(\lambda z) - f(z) = P_n(z)$$

is satisfied on the disk $\mathbb{D}$, then the functional equation

$$T_f(\lambda z) - T_f(z) = P_n(z)$$

is satisfied in some appropriate neighborhood in the exterior disk, where T_f is whatever other 'continuation' of f one wishes to consider. See [131] for a precise statement of this.

Gap series and pseudocontinuations were carried forward in two striking papers of A. B. Aleksandrov [**7, 8**] and in a moment we shall present a selection of his results, most of which come from [**8**]. Although Aleksandrov only works with pseudocontinuations, as we shall do here, some of his techniques apply to other forms of generalized analytic continuations. Also worth mentioning is how the application of Aleksandrov's criteria (see Theorem 6.9.14 below) in concrete circumstances leads to deep, in many cases unsolved, problems in number theory.

DEFINITION 6.9.13. Given a formal power series $\sum_n a_n z^n$ with $a_n \in \mathbb{C}$, its 'spectrum', denoted by Λ, is defined to be

$$\Lambda := \{n \in \mathbb{N} \cup \{0\} : a_n \neq 0\}.$$

If the spectrum of a formal power series is contained in a given subset of $\mathbb{N} \cup \{0\}$, we say the power series (or the function which it converges to, if convergent) is 'Λ-spectral'.

THEOREM 6.9.14 (Aleksandrov [**8**]). *Let f be an analytic function on $\mathbb{D}$ and pseudocontinuable across an open sub-arc $I \subset \mathbb{T}$ of length $|I|$. Suppose also that the Taylor expansion of f is Λ-spectral, where Λ posses the following property: There exists an $m \in \mathbb{N}$ and integers*

$$0 \leq r_1 < r_2 < \cdots < r_k \leq m - 1$$

such that

(6.9.15)
$$\frac{k}{m} < \frac{|I|}{2\pi}$$

and Λ is contained in the union of the k residue classes modulo m

$$r_j + m\mathbb{Z}, \quad j = 1, \ldots, k.$$

Then f is pseudocontinuable across $\mathbb{T}$ or, what is the same, for some $R > 1$ there is a meromorphic function g on the annulus $\{1 < |z| < R\}$ such that at almost every point of $\mathbb{T}$, f and g have equal non-tangential boundary values.

Before we proceed to the proof, let us illustrate the theorem.

COROLLARY 6.9.16. *The conclusion of the above theorem holds for an arbitrary arc $I \subset \mathbb{T}$, however small, if Λ is the set of squares*

$$S := \{0, 1, 4, 9, 16, \ldots\}.$$

PROOF. To deduce the corollary, we have to show that S satisfies the condition of Theorem 6.9.14 for pairs k, m with k/m as small as we please.

For any odd prime p, it is well known that among the integers $\{1, 2, \ldots, p - 1\}$, precisely half of them are quadratic residues modulo p [**104**]. If we choose the residue classes modulo p corresponding to these quadratic residues classes and adjoin the residue class $p\mathbb{Z}$, we have $(p+1)/2$ residue classes modulo p, whose union contains $S = \{0, 1, 4, 9, \ldots\}$. Let us denote these residue classes by

$$A_1, A_2, \ldots, A_{(p+1)/2}.$$

Now, if q is another odd prime, distinct from p, we can apply the above reasoning to show that S is also contained in the union of $(q + 1)/2$ residue classes modulo q, which we denote by

$$B_1, B_2, \ldots, B_{(q+1)/2}.$$

Thus

$$S \subset \left\{ \bigcup_{i=1}^{(p+1)/2} A_i \right\} \cap \left\{ \bigcup_{j=1}^{(q+1)/2} B_j \right\} = \bigcup_{i=1}^{(p+1)/2} \bigcup_{j=1}^{(q+1)/2} (A_i \cap B_j).$$

By the Chinese remainder theorem, each $A_i \cap B_j$ is a residue class modulo pq and so S is contained in the union of

$$\frac{p+1}{2} \frac{q+1}{2}$$

residue classes modulo pq.

We can repeat this procedure with new primes. Thus, choosing l distinct odd primes $p_1, p_2, \ldots, p_l$, we see that S is contained in a union of k distinct residue classes modulo m, where $m = p_1 p_2 \cdots p_l$ and

$$k = \frac{p_1+1}{2} \frac{p_2+1}{2} \cdots \frac{p_l+1}{2} \cdots \leq (2/3)^l p_1 p_2 \cdots p_l = (2/3)^l m.$$

For l sufficiently large, $k/m \leq (2/3)^l$ will be a small as we please. □

PROOF OF THEOREM 6.9.14. Since both the hypothesis and conclusion are invariant with respect to rotation, we may assume, without loss of generality, that

$$I = \{e^{i\theta} : 0 < \theta < \theta_0\}, \quad \theta_0 < 2\pi.$$

Assume f satisfies the hypothesis of the theorem. Thus the hypothesis holds with $2\pi k/m < \theta_0$. Let ε satisfy $0 < \varepsilon < 2\pi/m$. We shall show for ε sufficiently close to $2\pi/m$, that

(6.9.17) f has a pseudocontinuation across $\{e^{i\theta} : -\varepsilon < \theta < \theta_0\}$.

This implies the conclusion of the theorem, since after a finite number of iterations, it yields that f has a pseudocontinuation across all of $\mathbb{T}$.

Let $\omega = e^{2\pi i/m}$. Then the functions

$$f(\omega^j z), \quad j = 1, 2, \ldots, k$$

are all pseudocontinuable across the arc

$$J = \{e^{i\theta} : -\varepsilon < \theta < -\varepsilon + \frac{2\pi}{m}\}$$

(which intersects I) provided that

(6.9.18) $$-\varepsilon + \frac{2\pi}{m} + k\frac{2\pi}{m} < \theta_0 = |I|$$

which, in view of the assumption in eq.(6.9.15), is true for ε close enough to $2\pi/m$. We now fix ε so that the inequality in eq.(6.9.18) holds.

By assumption, we can write

(6.9.19) $$f = \sum_{j=1}^{k} F_j,$$

where the spectrum of F_j is a subset of the residue class $m\mathbb{Z} + r_j$. Hence, for $l = 1, 2, \cdots k$,

$$(6.9.20) \qquad f(\omega^l z) = \sum_{j=1}^{k} F_j(\omega^l z) = \sum_{j=1}^{k} \omega^{lr_j} F_j(z).$$

Looking at eq.(6.9.20) as a system of k linear equations to be solved for the F_j, we see that the corresponding matrix for this system is

$$[\omega^{lr_j}], \quad (l, j = 1, 2, \ldots, k),$$

and denoting $z_j = \omega^{r_j}$, the z_j are distinct and the determinant of the above matrix is of Vandermonde type and non-vanishing. Thus eq.(6.9.20) can be solved for the F_j, and so each F_j is expressible as a complex linear combination of the functions

$$f(\omega z), f(\omega^2 z), \ldots, f(\omega^k z).$$

Hence also, as we see from eq.(6.9.19), $f(z)$ is so expressible. Therefore, f is pseudocontinuable across J, and hence across $I \cup J$. This concludes the proof of eq.(6.9.17), and hence of the theorem. $\qquad \square$

REMARK 6.9.21. Obviously the proof works under the weaker assumption that Λ is covered by the union of residue classes, as in the assumption of Theorem 6.9.14, apart from some finite subset. The only change to the proof is that in eq.(6.9.20), there appears a polynomial $P_l(z)$ added onto the right side of the equation for $f(\omega^l z)$ so that $f(z)$ is a linear combination of $f(\omega z), f(\omega^2 z), \ldots, f(\omega^k z)$ plus some polynomial, which does not affect the conclusion that f is pseudocontinuable across J.

THEOREM 6.9.22 (Aleksandrov). *Suppose that f is analytic on $\mathbb{D}$ and its Taylor series is Λ-spectral, where Λ satisfies the following condition: For every positive integer s, there is a positive integer m and a residue class modulo m, say $m\mathbb{Z} + r$ (where $0 \le r \le m-1$), which contains $-s$, but at most finitely many elements of Λ. Then, if f has a pseudocontinuation across all of $\mathbb{T}$, it has an analytic continuation to $\{|z| < R\}$ for some $R > 1$.*

Again, before going to the proof, let us illustrate the use of the theorem.

COROLLARY 6.9.23. *If f is analytic on $\mathbb{D}$ and Λ-spectral, where Λ is the set of squares*

$$\{0, 1, 4, 9, \ldots\},$$

and has a pseudocontinuation across $\mathbb{T}$, then the Taylor series of f has a radius of convergence greater than one.

To prove this corollary, it suffices to verify the hypothesis of Theorem 6.9.22. This is a consequence of the following number theory proposition.

PROPOSITION 6.9.24. *Given $s \in \mathbb{N}$, the arithmetic progression*

$$\{-s + 4s^2 n : n \in \mathbb{Z}\}$$

is disjoint from $\{0, 1, 4, 9, \ldots\}$.

PROOF. Indeed suppose that for some non-negative integer k, $-s + 4s^2 n = k^2$, or equivalently, $s(4ns - 1) = k^2$. If s is a square, then $4ns - 1$ is also a square congruent to -1 modulo 4, a contradiction. If s is not a square, then there is a prime p and an odd integer r such that p^r divides s, while p^{r+1} does not divide s. Also note that $4ns - 1$ and s have no common factors and so k^2 is divisible by p^r but not by p^{r+1}, which is a contradiction of the oddness of r.   □

Combining the corollaries to Theorem 6.9.14 and Theorem 6.9.22 we deduce the following result.

COROLLARY 6.9.25. *If f is analytic on $\mathbb{D}$ and its spectrum is contained in the set of squares $\{0, 1, 4, 9, \ldots\}$ (with possibly a finite number of exceptions) and has radius of convergence equal to one, then f does not have a pseudocontinuation across any arc of $\mathbb{T}$.*

PROOF OF THEOREM 6.9.22. The proof is based on the following well-known identity whose proof is left to the reader: If $f(z) = \sum a_n z^n$ converges for $|z| < R$, then for any positive integer m and any integer r with $0 \le r \le m - 1$

$$(6.9.26) \qquad \frac{1}{m} \sum_{j=0}^{m-1} \omega^{-rj} f(\omega^j z) = \sum_{n \equiv r \,(\text{mod } m)} a_n z^n,$$

where $\omega := e^{2\pi i/m}$.

By assumption, there is some $R > 1$ and some g meromorphic on $\{1 < |z| < R\}$ which is a pseudocontinuation of f across $\mathbb{T}$. Choose numbers $1 \le R_1 < R_2 \le R$ such that g has no poles in the annulus $A := \{R_1 < |z| < R_2\}$. Thus, g can be expressed in A by a convergent Laurent series

$$g(z) = \sum_{n=-\infty}^{\infty} c_n z^n, \quad z \in A.$$

We consider two cases.

First suppose that $c_n = 0$ for all $n < 0$ (except possibly for a finite number of n). Then g has an analytic continuation to the punctured disk $\{0 < |z| < R_2\}$ and this extension must coincide with f on $\mathbb{D} \backslash \{0\}$, which implies the conclusion of the theorem. Now assume that $c_n \ne 0$ for an infinite number of negative n. Suppose that $c_{-s} \ne 0$ for some $s > 0$. Now choose m and r as in the statement of the theorem. Then, by our hypothesis and eq.(6.9.26),

$$\frac{1}{m} \sum_{j=0}^{m-1} \omega^{-rj} f(\omega^j z)$$

reduces to a polynomial $P(z)$. On the other hand, this sum admits

$$\frac{1}{m} \sum_{j=0}^{m-1} \omega^{-rj} g(\omega^j z)$$

as a pseudocontinuation across $\mathbb{T}$. Consequently, the latter sum is a pseudocontinuation of $P(z)$ and so it reduces to a polynomial. But, again by eq.(6.9.26) (which also applies to Laurent series),

$$\frac{1}{m}\sum_{j=0}^{m-1}\omega^{rj}g(\omega^j z) = \sum_{n\in\mathbb{Z},n\equiv r\,(\text{mod }m)} c_n z^n.$$

But the sum on the right hand side includes the term $c_{-s}z^{-s}$ and so it can not be a polynomial. This is a contradiction and the proof is complete.   □

REMARK 6.9.27. It is clear from the proof that it would have been enough to assume that the hypotheses of the theorem hold, not for all positive s, but for all outside some finite set.

We end this section with a remark that in [8] Aleksandrov proves that other 'non-standard' gap series with radius of convergence equal to one, such as

$$\sum_{p\in P} c_p z^p,$$

where P is the set of primes, do not have pseudocontinuations across any arc of $\mathbb{T}$. Similarly, the gap series (when it has radius of convergence equal to one)

$$\sum_{b\in B_s} c_b z^b,$$

where B_s is the set of integers representable as a product of at most s prime numbers, does not have a pseudocontinuation across any arc of $\mathbb{T}$.

6.10. Functional equations and non-continuability

Starting from the earliest examples of Weierstrass, a basic theme in constructing non-continuable power series (and later non-continuable Dirichlet series, as well as series not amenable to generalized methods of continuation) is the use of 'functional equations'. For example, if

$$(6.10.1) \qquad f(z) = \sum_{n=0}^{\infty} c_n z^{2^n}$$

has radius of convergence equal to one, its non-continuability across any point of the unit circle is an immediate consequence of the fact that for each $m \in \mathbb{N}$, f satisfies the functional equation

$$(6.10.2) \qquad f(\omega_{2^m} z) = f(z) + P_m(z), \;\; z \in \mathbb{D},$$

where

$$\omega_{2^m} = e^{2\pi i/2^m}$$

and P_m is a suitable polynomial. The point is, if f were analytically continuable across an arc I of $\mathbb{T}$, the functional equation in eq.(6.10.2) 'squeezes through' I for all sufficiently large m, and so continues to hold on a portion of $\mathbb{D}_e$ contiguous to I, forcing the analyticity of f to propagate to a full neighborhood of $\mathbb{T}$. This analytic continuation to an open disk properly containing $\mathbb{D}^-$ violates the hypothesis that the radius of convergence of f is equal to one.

As we have seen (Theorem 6.9.7), with very little modification, the above argument also shows f has no pseudocontinuation across any arc I of $\mathbb{T}$. Jöricke [80, 81] uses similar ideas to construct gap series which are not pseudocontinuable across any set of *positive measure*[6] in $\mathbb{T}$. The same idea underlies Aleksandrov's examples of 'mildly lacunary' Taylor series without pseudocontinuations, which were presented in § 6.9 of this chapter.

An essentially new, and very fruitful idea of this kind was observed by J. F. Ritt in his doctoral thesis [116]. To describe Ritt's results, we bring in linear differential equations of infinite order as studied in § 6.8 of this chapter. Suppose that ϕ is a function of the class, we henceforth shall denote by Z, of entire functions of exponential type zero, that is to say, the condition

$$|\phi(w)| = O(e^{\varepsilon|w|}), \quad |w| \to \infty$$

holds for each $\varepsilon > 0$. Then it is fairly easy to check that for f analytic on a disk G, the function $\phi(D)f$, as defined earlier (see Definition 6.8.3) by

$$(\phi(D)f)(z) = \sum_{j=0}^{\infty} b_j f^{(j)}(z),$$

where $\{b_j\}$ are the Taylor coefficients of ϕ, is well defined for each $z \in G$ and moreover, the series converges uniformly on compact subsets of G. Since the radius of convergence of G may be arbitrarily small, the map $f \to \phi(D)f$, for any fixed $\phi \in Z$, is a local operator on germs of analytic functions. In particular, if f is analytically continuable along any arc in $\mathbb{C}$ (which is permitted to intersect itself), the same is true for $\phi(D)f$. Thus, in this sense, $\phi(D)$ is like an ordinary differential operator and $\phi(D)f$ is uniquely defined on the whole Riemann surface of f.

What we have said so far was in essence known before Ritt (see his references to Pincherle and Bourlet). These ideas must have been in the air around 1917, because less than a year after Ritt's thesis was completed, H. Cramér [39] in his doctoral dissertation in Stockholm independently studied the operators $\phi(D)f$, where ϕ was of (not necessarily zero) exponential type. However, Ritt made a striking application of this idea to non-continuability of Dirichlet series which we discuss in a moment. We remark that Ritt's work required a somewhat stronger hypothesis on ϕ, namely $\phi \in Z_0$, where by Z_0 we mean the class of entire functions of 'genus zero', that is functions of the form

$$(6.10.3) \qquad\qquad \phi(w) = Cw^r \prod_{n=1}^{\infty}\left(1 - \frac{w}{w_n}\right),$$

where C is a constant, $r \in \mathbb{N} \cup \{0\}$, and $\{w_n\}$ is a sequence of (not necessarily distinct) non-zero complex numbers satisfying

$$\sum_{n=1}^{\infty} \frac{1}{|w_n|} < \infty.$$

The reader may easily verify that $Z_0 \subset Z$ (or consult [23] for more details, where the notation, however, is different).

---

[6]The result is the following: Suppose that $n_{k+1} = \alpha_k n_k$, where $\{n_k\}$ and $\{\alpha_k\}$ are sequences of integers with $n_k \geq 1$ and $\alpha_k \geq 2$. If the power series $f = \sum_k a_k z^{n_k}$ has radius of convergence equal to one, then f does not have a 'pseudocontinuation across any set of positive measure' in $\mathbb{T}$. Here we say f has a 'pseudocontinuation across a set $E \subset \mathbb{T}$ of positive measure' if (i) there is an open set $\Omega \subset \mathbb{D}_e$ such that $\Omega^- \cap \mathbb{T} \supset E$; (ii) at each point $e^{i\theta} \in E$, the set Ω contains a truncated non-tangential cone $K_{e^{i\theta}}$ with vertex at $e^{i\theta}$; (iii) there is an $F \in \mathfrak{H}(\Omega)$ such that for every $e^{i\theta} \in E$, $\lim_{z \to e^{i\theta}, z \in K_{e^{i\theta}}} F(z)$ is equal to the non-tangential limit (as $z \to e^{i\theta}$) of f.

Consider now a function f which is analytic on a disk G and satisfies there, for some non-trivial $\phi \in Z_0$,

$$\phi(D)f = 0.$$

For example, in eq.(6.10.3), let $r = 0$ and all the w_n's be distinct. Then any generalized Dirichlet series[7]

$$(6.10.4) \qquad\qquad f(z) := \sum_{n=1}^{\infty} A_n e^{w_n z},$$

which also converges uniformly on compact subsets of G, satisfies $\phi(D)f = 0$ on G. If there are higher multiplicity w_n's, we can include in eq.(6.10.4) terms involving

$$z^j e^{w_n z}, \quad j = 1, 2, \ldots, k_n - 1,$$

where k_n is the multiplicity of w_n. We leave this extension to the reader, our purpose being to illustrate the main points of Ritt's work. By what has been said, if f is continuable along any path, the function $\phi(D)f$ is as well and if $\phi(D)f = 0$ holds initially, this will persist. Ritt [116] proved the following striking results.

THEOREM 6.10.5 (Ritt). *Suppose f is analytic on a disk G, ϕ is a non-trivial function in Z_0, and f satisfies $\phi(D)f = 0$ on G. If f is analytically continuable along any path in $\mathbb{C}$ from a point $z_0 \in G$ back to z_0, then this continuation of f coincides in a neighborhood of z_0 with the initial branch.*

Roughly speaking: under the stated hypothesis, analytic continuations of f satisfying $\phi(D)f = 0$ must remain single-valued.

THEOREM 6.10.6 (Ritt). *Under the same hypothesis, if f is analytically continuable from z_0 back to z_0 along a circular arc γ, it extends analytically throughout the disk bounded by γ.*

Thus, not only are there no winding points enclosed by γ, but there are no singularities whatsoever. Later, Ritt [117] (and independently Hille [74]) remarked that the circular arc γ in the previous result could be permitted to enclose an arbitrary Jordan domain.

Ritt also showed, on the basis of Theorem 6.10.6, that a given Dirichlet series of the form

$$F(z) = \sum_{m=1}^{\infty} A_n e^{w_m z},$$

where

$$w_m > 0 \quad \text{and} \quad \sum_{n=1}^{\infty} \frac{1}{w_m} < \infty,$$

which converges uniformly on compact subsets of the half-plane $L := \{z : \Re z < 0\}$, but does not do so for any half-plane $\{z : \Re z < \varepsilon\}$ with $\varepsilon > 0$, is not analytically continuable across any boundary point of L. This implies the following 'gap

---

[7]We use the term 'generalized' here since Dirichlet series, as they are presented in most textbooks require the w_n's to be positive and strictly increasing to infinity (or negative and strictly decreasing towards minus infinity).

theorem': if

$$g(z) = \sum_{j=0}^{\infty} a_j z^{n_j},$$

where

$$n_j \in \mathbb{N} \quad \text{with} \quad \sum_{j=1}^{\infty} \frac{1}{n_j} < \infty,$$

has radius of convergence equal to one, then g is not analytically continuable across any point of $\mathbb{T}$. Indeed, this follows by applying the Ritt result just mentioned to the function

$$(6.10.7) \qquad\qquad F(z) := g(e^z) = \sum_{j=0}^{\infty} a_j e^{n_j z}, \quad z \in L.$$

We shall not reproduce Ritt's results here, since they were sharpened in an essential way by Pólya [**107**] who showed the following.

THEOREM 6.10.8 (Pólya). *Suppose ϕ is a non-trivial function from Z and f is analytic on a disk G satisfying $\phi(D)f = 0$. If f is analytically continuable from a point $z_0 \in G$ back to $z_0 \in G$ along any Jordan arc γ, then f extends analytically to a neighborhood of the convex hull of γ, or roughly expressed: the maximal domain of existence of any solution f of $\phi(D)f = 0$ is convex.*

This truly remarkable result extends those of Ritt in two essential ways (i) allowing ϕ to be in the more natural class Z (rather than Z_0), and (ii) yielding the convexity of the domain of existence (whereas Ritt could obtain only simple connectedness).

As Pólya went on to show, Theorem 6.10.8, together with an elementary result concerning the zero sets of entire functions in the class Z, yields the full Fabry gap theorem (Theorem 6.9.4). Indeed, suppose

$$g(z) = \sum_{j=0}^{\infty} a_j z^{n_j}$$

has radius of convergence equal to one and $n_j/j \to \infty$. In order to prove that g does not have an analytic continuation across any point of $\mathbb{T}$, one again introduces the function F in eq.(6.10.7). The crucial step is to show that $\phi(D)F = 0$ on the half-plane L for a non-trivial $\phi \in Z$. This will certainly hold if $\phi(n_j) = 0$ for all j. Thus the missing step is contained in the following proposition.

PROPOSITION 6.10.9. *If $\{n_j : j = 1, 2, \ldots\}$ are positive integers with $n_j/j \to \infty$, there is a non-trivial $\phi \in Z$ vanishing on $\{n_j\}$.*

PROOF. The function

$$\phi(w) := \prod_{j=1}^{\infty} \left(1 - \frac{w^2}{n_j^2}\right)$$

is, under the stated conditions on $\{n_j\}$, a non-trivial entire function, the product converging uniformly on compact sets. To estimate ϕ, we use the well-known formula

$$\frac{\sin(\pi z)}{\pi z} = \prod_{n=1}^{\infty} (1 - \frac{z^2}{n^2}).$$

Putting here $z = it$ with $t > 0$ gives us

(6.10.10)
$$\prod_{n=1}^{\infty} (1 + \frac{t^2}{n^2}) = \frac{\sinh(\pi t)}{\pi t} \le C e^{\pi t}.$$

Now, for each positive integer N,

$$|\phi(w)| \le (1 + |w|^2)^N \prod_{j>N} (1 + \frac{|w|^2}{n_j^2}).$$

For any given $\lambda > 0$, $n_j \ge \lambda j$ as soon as $j > N(\lambda)$ and we get, using eq.(6.10.10),

$$|\phi(w)| \le (1 + |w|^2)^{N(\lambda)} \prod_{j>N(\lambda)} (1 + \frac{|w|^2}{\lambda^2 j^2}) \le (1 + |w|^2)^{N(\lambda)} C e^{\pi |w|/\lambda}.$$

Hence

$$\varlimsup_{|w|\to\infty} |\phi(w)| e^{-c|w|} = 0$$

for every $c > \pi/\lambda$. Since λ may be chosen arbitrarily large, we conclude that $\phi \in Z$. $\square$

In closing this section, we remark that another very elegant proof of Fabry's gap theorem, based on interpolation by entire functions, is given in [**23**, Sect. 12.6]. Another nice treatment of the Fabry gap theorem can be found in a book of Havin and Jöricke [**69**]. The methods based on interpolation and on functional equations have proven very useful in more recent multivariable generalizations. For some orientation and further references, the reader may consult [**88**].

CHAPTER 7

A continuation involving almost periodic functions

7.1. Revisiting the Poincaré example

Recall the example of Poincaré (Theorem 3.1.1) involving the function

$$p(z) = \sum_{n=1}^{\infty} \frac{c_n}{1 - e^{i\theta_n} z},$$

where $\{e^{i\theta_n}\}$ was a sequence of distinct points dense in $\mathbb{T}$, and $\{c_n\}$ was an absolutely summable sequence of non-zero complex numbers. Poincaré showed that the unit circle $\mathbb{T}$ is a natural boundary for $f := p|\mathbb{D}$ in the sense that f does not have an analytic continuation across any point of $\mathbb{T}$. A similar result is true for $F := p|\mathbb{D}_e$. However, we know from more modern techniques (see Chapter 3), not available to Poincaré, that f and F are pseudocontinuations of each other in the sense that they have equal non-tangential limits for almost every $e^{i\theta}$.

The Taylor expansions for f and F are

$$f(z) = \sum_{n=0}^{\infty} A(n) z^n, \quad z \in \mathbb{D},$$

$$F(z) = -\sum_{n=1}^{\infty} \frac{A(-n)}{z^n}, \quad z \in \mathbb{D}_e,$$

where the coefficients $\{A(n)\}$ are given by the formula

(7.1.1)
$$A(n) = \sum_{m=1}^{\infty} c_m e^{in\theta_m}, \quad n \in \mathbb{Z}.$$

Our point of view so far was to interpret the sequence $\{A(n)\}$ as the Fourier coefficients of the purely atomic measure

$$d\mu = \sum_{m=1}^{\infty} c_m \delta_{e^{-i\theta_m}},$$

that is to say

$$A(n) = \int e^{-int} d\mu(e^{it}).$$

However, the formula in eq.(7.1.1) also suggests another line of generalization. Since $n \to A(n)$ is an almost periodic function on $\mathbb{Z}$ (see Definition 7.2.8 below), one might wonder about the coherence properties of the functions f and F when the Fourier sequence $\{A(n)\}$ is replaced by an almost periodic sequence.

Define $\mathfrak{A}$ to be the class of analytic functions of the form

$$f_A(z) = \sum_{n=0}^{\infty} A(n)z^n,$$

where A is an almost periodic function from $\mathbb{Z}$ to $\mathbb{C}$. Since $\{A(n)\}$ is a bounded sequence, the above series converges in $\mathbb{D}$. Likewise, the candidate to be its 'continuation', the function

$$F_A(z) = -\sum_{n=1}^{\infty} \frac{A(-n)}{z^n},$$

converges in $\mathbb{D}_e$. Let $\widetilde{\mathfrak{A}}$ denote the class of such F_A.

There are a few things to notice about these two functions. First, the extra minus sign is needed in the definition of F_A since if A were the most basic type of almost periodic sequence

$$A(n) = \sum_{k=1}^{K} a_k e^{in\theta_k}, \quad n \in \mathbb{Z},$$

then a power series computation shows that both f_A and F_A are equal to the same rational function, namely

$$\sum_{k=1}^{K} \frac{a_k}{1 - e^{i\theta_k} z}.$$

Secondly, f_A and F_A enjoy some 'coherence' properties since they uniquely determine each other. This comes from the fact that an almost periodic function on $\mathbb{Z}$ is uniquely determined by its values on $\mathbb{Z}_+$ (or $\mathbb{Z}_-$), see Proposition 7.2.13 below. Thirdly, there is no reason to believe that functions in $\mathfrak{A}$ possess non-tangential limit values almost everywhere [1] and so the mapping $f_A \to F_A$ from $\mathfrak{A}$ to $\widetilde{\mathfrak{A}}$ does not seem to be a correspondence mediated by a pseudocontinuation. However, as we shall see below, after some appropriate machinery is developed, the correspondence between f_A and F_A is compatible with analytic continuation, that is to say, an Abelian theorem exists for this type of 'continuation'.

Results of this type have been investigated by Bochner and Bohnenblust [**24**] and in fact, there is even a continuous version of this Abelian theorem where the functions f_A and F_A above are replaced by the Laplace transforms

$$f_\phi(z) = \int_0^{\infty} \phi(t)e^{-tz}dt, \quad z = x + iy, \quad x > 0$$

$$F_\phi(z) = -\int_{-\infty}^{0} \phi(t)e^{-tz}dt, \quad z = x + iy, \quad x < 0$$

of a (Bohr) almost periodic function ϕ (see below) on the real line $\mathbb{R}$.

---

[1] We do not have a proof of this fact but our intuition says that one should be able to create an example of an almost periodic sequence $\{A(n)\}$ such that f_A does not have non-tangential limits almost everywhere.

7.2. A primer on almost periodic functions

We now quickly review some basic facts about almost periodic functions and refer the reader to some standard monographs on the subject for further details [**25, 38, 87, 96, 97, 105**]. For a function $f \in BUC(\mathbb{R})$ (complex-valued, bounded, uniformly continuous functions on $\mathbb{R}$ - endowed with the sup-norm topology), let

$$f_y(x) = f(x - y), \;\; y \in \mathbb{R},$$

be the translates of f. For $\lambda \in \mathbb{R}$, let

$$e_\lambda(x) = e^{i\lambda x}, \;\; x \in \mathbb{R}.$$

PROPOSITION 7.2.1. *For a function $f \in BUC(\mathbb{R})$ the following two conditions are equivalent:*

(1) $\{f_y : y \in \mathbb{R}\}^-$ *is compact in $BUC(\mathbb{R})$.*

(2) f *is the sup-norm limit of finite linear combinations of the functions*

$$\{e_\lambda : \lambda \in \mathbb{R}\}.$$

DEFINITION 7.2.2. *A function $f \in BUC(\mathbb{R})$ satisfying either of the equivalent conditions of Proposition 7.2.1 is said to be Bohr 'almost periodic' (after H. Bohr [**25**]). We denote this class by $AP(\mathbb{R})$.*

It turns out that $AP(\mathbb{R})$ is a norm closed subalgebra of $BUC(\mathbb{R})$.

THEOREM 7.2.3 (Bohr [**25**]). *For each $f \in AP(\mathbb{R})$, the following limit*

$$M(f) := \lim_{T \to \infty} \frac{1}{2T} \int_{-T}^{T} f(x) dx$$

exists and is called the 'Bohr mean value' of f.

We point out that $M(f_y) = M(f)$, $M(f) \geq 0$ whenever $f \geq 0$, and $M(1) = 1$.

DEFINITION 7.2.4. *For $f \in AP(\mathbb{R})$ and $\lambda \in \mathbb{R}$, note that $e_{-\lambda}f \in AP(\mathbb{R})$. The 'Bohr spectrum' of f is defined to be the set*

$$\Omega(f) := \left\{ \lambda \in \mathbb{R} : M(e_{-\lambda}f) \neq 0 \right\}.$$

THEOREM 7.2.5 (Bohr [**25**]). *If f is a non-trivial almost periodic function, then $\Omega(f)$ is non-empty and moreover, $\Omega(f)$ is at most a countable (but not necessarily closed) set of real numbers.*

For example, if

$$f(t) = \sum_{k=1}^{K} a_k e^{i\lambda_k t},$$

then

$$M(e_{-\lambda}f) = \begin{cases} 0 & \text{if } \lambda \neq \lambda_k, \;\; k = 1, \ldots, K, \\ a_k & \text{if } \lambda = \lambda_k. \end{cases}$$

and so

$$\Omega(f) = \{\lambda_1, \ldots, \lambda_K\}.$$

REMARK 7.2.6.   (1) If $f \in AP(\mathbb{R})$ and $\Omega(f) = \{\lambda_k\}$, then we can think of f as having a 'Fourier' expansion

$$f \sim \sum_{k=1}^{\infty} a_k e^{i\lambda_k x},$$

where $a_k = M(e_{-\lambda_k} f)$. Moreover, we also have a 'Parseval's equality' [**38**, p. 28]

$$\sum_{k=1}^{\infty} |a_k|^2 = M(|f|^2).$$

From this follows that two distinct AP functions have distinct Fourier expansions.

(2) Also [**38**, p. 41], there exists trigonometric polynomials

$$\sigma_m(x) = \sum_{k=1}^{n_m} r_{k,m} a_k e^{i\lambda_k x}$$

which converge uniformly to f as $m \to \infty$. The numbers $r_{k,m}$ are rational and depend on λ_k (the 'Fourier exponents' of f) and m, but not on a_m. Thus, not only can an almost periodic function f be approximated by a sequence of trigonometric polynomials, but it can be approximated by trigonometric polynomials $\{\sigma_m\}$ satisfying $\Omega(\sigma_m) \subset \Omega(f)$.

We now define the set $AP(\mathbb{Z})$ of 'almost periodic sequences'. Let $\ell^\infty(\mathbb{Z})$ denote the set of bounded two-sided sequences of complex numbers with the norm

$$\|A\| = \sup\{|A(n)| : n \in \mathbb{Z}\}.$$

PROPOSITION 7.2.7. *For $A \in \ell^\infty(\mathbb{Z})$, following are equivalent:*

(1) *The set*

$$\left\{ \{A(\cdot - m)\} : m \in \mathbb{Z} \right\}^-$$

is compact in $\ell^\infty(\mathbb{Z})$.

(2) *A is the norm limit of finite linear combinations of the characters*

$$\{n \to e^{in\lambda} : \lambda \in [0, 2\pi)\}.$$

(3) *There exists an $f \in AP(\mathbb{R})$ such that $f(n) = A(n)$ for all $n \in \mathbb{Z}$.*

DEFINITION 7.2.8. The class of sequences in $\ell^\infty(\mathbb{Z})$ satisfying one of the above equivalent conditions is called the 'almost periodic sequences' and is denoted by $AP(\mathbb{Z})$.

Almost exactly as we did for $AP(\mathbb{R})$ functions, we can define the 'Bohr mean value' of $A \in AP(\mathbb{Z})$ by

$$M(A) := \lim_{m \to \infty} \frac{1}{2m+1} \sum_{k=-m}^{m} A(k)$$

and the 'Bohr spectrum' of A by

(7.2.9) $\Omega(A) := \left\{ e^{i\theta} : M_\theta(A) := \lim_{m \to \infty} \frac{1}{2m+1} \sum_{k=-m}^{m} A(k) e^{-ik\theta} \neq 0 \right\}.$

REMARK 7.2.10. In fact [**38**, p. 48], if $f \in AP(\mathbb{R})$ such that $f(n) = A(n)$ for all $n \in \mathbb{Z}$ (Proposition 7.2.7), then $M(A) = M(f)$ (as in Theorem 7.2.5). Thus $\Omega(f) = \Omega(A)$.

As before, $\Omega(A)$ is non-empty and at most countable for a non-trivial $A \in AP(\mathbb{Z})$. For example, if $A \in AP(\mathbb{Z})$ is a finite linear combination of the characters,

$$n \to e^{in\theta_k}, \quad k = 1, \ldots, K, \quad \theta_k \in [0, 2\pi),$$

that is to say,

$$A(n) = \sum_{k=1}^{K} a_k e^{in\theta_k}, \quad n \in \mathbb{Z},$$

then

(7.2.11)
$$M_\theta(A) = \begin{cases} 0 & \text{if } \theta \neq \theta_k, \quad k = 1, \ldots, K, \\ a_k & \text{if } \theta = \theta_k. \end{cases}$$

and so

$$\Omega(A) = \{e^{i\theta_k} : k = 1, \ldots, K\}.$$

REMARK 7.2.12. By Remark 7.2.6 and Remark 7.2.10, an $A \in AP(\mathbb{Z})$ can be approximated, in the norm of $\ell^\infty(\mathbb{Z})$, by a sequence $\{A_n\}$, each a finite linear combination of characters, such that $\Omega(A_n) \subset \Omega(A)$.

One of the important reasons we are looking at the coherence properties of the functions f_A and F_A, where $A \in AP(\mathbb{Z})$ is that they uniquely determine each other in the sense that if $f_A \equiv 0$, *i.e.*, $A(n) = 0$ for all $n \geq 0$, then F_A must also be the zero function, *i.e.*, $A(n) = 0$ for all $n \leq 0$. This follows from the following basic fact about almost periodic sequences.

PROPOSITION 7.2.13. *If $A \in AP(\mathbb{Z})$ and $A(n) = 0$ for all $n \geq 0$, then $A(n) = 0$ for all $n \in \mathbb{Z}$.*

PROOF. Suppose that $A(n) = 0$ for $n \geq 0$ but $A(-1) \neq 0$. Then for $K, L \in \mathbb{N} \cup \{0\}, K \neq L$,

$$\sup\{|A(n - K) - A(n - L)| : n \in \mathbb{Z}\} \geq |A(-1)| \neq 0$$

and so the sequence

$$\big\{ \{A(\cdot - K)\} : K \geq 0 \big\}$$

does not have a convergent subsequence in $\ell^\infty(\mathbb{Z})$, contradicting the compactness condition (2) of Proposition 7.2.7. In a similar way, one can show that $A(-2), A(-3), \ldots$ are all equal to zero as well. □

7.3. Compatibility with analytic continuation

Now that we have defined a 'continuation' $f_A \to F_A$ via almost periodic sequences, our first task, as was our first task with all our other types of continuations (Borel-Walsh, Gončar, pseudocontinuation), will be to show the following 'Abelian theorem', making the continuation $f_A \to F_A$ compatible with analytic continuation.

THEOREM 7.3.1. *Suppose $A \in AP(\mathbb{Z})$ and*

$$f_A(z) = \sum_{n=0}^{\infty} A(n)z^n, \quad z \in \mathbb{D},$$

$$F_A(z) = -\sum_{n=1}^{\infty} \frac{A(-n)}{z^n}, \quad z \in \mathbb{D}_e.$$

If f_A has an analytic continuation across some boundary arc $J \subset \mathbb{T}$, then this analytic continuation must be equal to F_A.

We begin our proof with the following lemma which is reminiscent of Proposition 3.2.2. Recall the definition of $M_\theta(A)$ from eq.(7.2.9).

LEMMA 7.3.2. *For $A \in AP(\mathbb{Z})$ and f_A as above,*

$$M_{-\theta}(A) = \lim_{r \to 1^-} (1-r)f_A(re^{i\theta}) \quad \text{for all } \theta \in [0, 2\pi).$$

PROOF. Let $\varepsilon > 0$ and choose a finite linear combination of the characters $\{n \to e^{in\lambda} : \lambda \in [0, 2\pi)\}$, i.e.,

$$B(n) = \sum_{k=1}^{K} b_k e^{in\theta_k}, \quad n \in \mathbb{Z},$$

with $\|A - B\| < \varepsilon/2$ (Proposition 7.2.7). Observe two basic facts,

(7.3.3)
$$f_B(z) = \sum_{k=1}^{K} \frac{b_k}{1 - e^{i\theta_k}z}$$

(7.3.4) $|M_\theta(A) - M_\theta(B)| \le \|A - B\| < \varepsilon/2$ for all $\theta \in [0, 2\pi)$.

Write $f_A(re^{i\theta})$ as

$$f_A(re^{i\theta}) = f_B(re^{i\theta}) + \left[f_A(re^{i\theta}) - f_B(re^{i\theta}) \right]$$

and note that from eq.(7.3.3) and the following estimate (from the definitions of f_A and f_B)

$$(1-r)|f_A(re^{i\theta}) - f_B(re^{i\theta})| \le (1-r) \sum_{n=0}^{\infty} |A(n) - B(n)|r^n \le \|A - B\| \le \varepsilon/2,$$

we obtain

$$\left| (1-r)f_A(re^{i\theta}) - M_{-\theta}(A) \right| \le \left| (1-r) \sum_{k=1}^{K} \frac{b_k}{1 - e^{i\theta_k}re^{i\theta}} - M_{-\theta}(A) \right| + \varepsilon/2.$$

Noticing from eq.(7.2.11) that

$$\lim_{r \to 1^-} (1-r) \sum_{k=1}^{K} \frac{b_k}{1 - e^{i\theta_k}re^{i\theta}} = M_{-\theta}(B),$$

we now get

$$\overline{\lim_{r \to 1^-}} \left| (1-r)f_A(re^{i\theta}) - M_{-\theta}(A) \right| \le |M_{-\theta}(B) - M_{-\theta}(A)| + \varepsilon/2,$$

which, by eq.(7.3.4), is bounded above by ε. The result now follows. $\square$

REMARK 7.3.5. This result implies

$$|f_A(re^{i\theta})| \to +\infty \quad \text{as } r \to 1^-$$

for each $e^{i\theta} \in \Omega(A)$.

PROOF OF THEOREM 7.3.1. First note that if f_A has an analytic continuation across some boundary arc $J \subset \mathbb{T}$, then for each $e^{i\theta} \in J$, $f_A(re^{i\theta})$ remains bounded as $r \to 1^-$. Thus by Lemma 7.3.2,

$$e^{-i\theta} \notin \Omega(A) \quad \text{for all } e^{i\theta} \in J.$$

This means, by Remark 7.2.12, that A can be approximated by a sequence $\{A_s : s = 1, 2, \ldots\}$ of $AP(\mathbb{Z})$ sequences of the form

$$A_s(n) = \sum_{k=1}^{K_s} a_{s,k} e^{in\theta_{s,k}}, \quad n \in \mathbb{Z},$$

where

$$e^{-i\theta_{s,k}} \notin J.$$

In fact, we can arrange $\|A - A_s\| \le 1/s$, where $\|\cdot\|$ denotes the uniform norm. A computation, as before, with power series shows that

$$f_{A_s}(z) = \sum_{k=1}^{K_s} \frac{a_{s,k}}{1 - e^{i\theta_{s,k}}z} := R_s(z), \quad |z| < 1$$

and $F_{A_s}(z) = R_s(z), |z| > 1$. Note that

$$|f_A(z) - R_s(z)| \le \sum_{n=0}^{\infty} |A(n) - A_s(n)||z|^n \le \frac{1}{s} \frac{1}{1 - |z|}, \quad |z| < 1,$$

$$|F_A(z) - R_s(z)| \le \frac{1}{s} \frac{1}{|z| - 1}, \quad |z| > 1.$$

Finally, if γ is an circle with center in J which does not contain either of the endpoints of J, then, from the previous two inequalities,

$$|R_s(z)| \le \frac{C}{\big| |z| - 1 \big|}, \quad z \in \text{int}(\gamma),$$

where C is a positive constant independent of s. Thus the sequence $\{R_s\}$ is a normal family on the interior of γ (see Lemma 7.3.6 below) and so there is a subsequence converging uniformly on compact subsets of the interior of γ. From the above two estimates, F_A is an analytic continuation of f_A across J. □

Needed in the above proof is the following lemma which, in a sense, is related to a theorem of Beurling [48]. We found this proof in a paper of Sundberg [136]. We will see a result like this used again (see Theorem 8.2.2).

LEMMA 7.3.6. *Let U be an open set in $\mathbb{C}$ and $\mathcal{F}$ be a family of functions from $\mathfrak{H}(U)$. If there is a $\rho \in L^1(U, dA)$ such that*

$$\log^+ |f(z)| \le \rho(z), \quad \text{for all } f \in \mathcal{F} \text{ and } z \in U,$$

then $\mathcal{F}$ is a normal family.

PROOF. By Montel's theorem, it suffices to show that given any compact set $K \subset U$,

$$\sup\{|f(z)| : z \in K, f \in \mathcal{F}\} < \infty.$$

To this end, fix a compact $K \subset U$ and pick a $\delta > 0$ such that $K_\delta := \{z \in \mathbb{C} : \operatorname{dist}(z, K) \leq \delta\} \subset U$. Then if $f \in \mathcal{F}$ and $z \in K$, the subharmonicity of $\log^+|f|$ implies that

$$\log^+|f(z)| \leq \frac{1}{\pi\delta^2} \int_{|w-z| \leq \delta} \log^+|f(w)| dA(w) \leq \frac{1}{\pi\delta^2} \int_{K_\delta} \rho(w) dA(w).$$

Thus

$$|f(z)| \leq \exp\left[\frac{1}{\pi\delta^2} \int_{K_\delta} \rho(w) dA(w)\right] \quad \text{for all} \in \mathcal{F} \text{ and } z \in K.$$

$\square$

We now discuss a continuous version of Theorem 7.3.1 where the functions f_A and F_A are replaced by the Laplace transforms

$$f_\phi(z) = \int_0^\infty \phi(t) e^{-tz} dt, \quad z = x + iy, \ x > 0$$

$$F_\phi(z) = -\int_{-\infty}^0 \phi(t) e^{-tz} dt, \quad z = x + iy, \ x < 0,$$

where $\phi \in AP(\mathbb{R})$. Like in the discrete case, f_ϕ and F_ϕ uniquely determine each other in the sense that $f_\phi \equiv 0$ implies $\phi|\mathbb{R}_+ = 0$, which, by almost periodicity (exactly as in the discrete case) yields $\phi|\mathbb{R}_- = 0$ and so $F_\phi \equiv 0$. Our main theorem here is that the 'continuation' $f_\phi \to F_\phi$ is compatible with analytic continuation.

THEOREM 7.3.7. *Suppose* $\phi \in AP(\mathbb{R})$ *and* f_ϕ *and* F_ϕ *are defined as above. If* f_ϕ *has an analytic continuation across some interval* $(ia, ib) \subset i\mathbb{R}$, *then this continuation must be equal to* F_ϕ.

Before starting the proof, we need the following preliminary fact which is the continuous analog to Lemma 7.3.2.

LEMMA 7.3.8. *For* $\phi \in AP(\mathbb{R})$,

$$\lim_{x \to 0^+} x f_\phi(x + iy) = M(e_{-y}\phi).$$

PROOF. Let $\varepsilon > 0$ be given and let

$$\phi_n(t) = \sum_{k=1}^K a_{n,k} e^{it\lambda_{n,k}}$$

be such that $|\phi_n(t) - \phi(t)| < \varepsilon$ for all $t \in \mathbb{R}$. A computation shows that

$$f_{\phi_n}(z) = \sum_{k=1}^K \frac{a_{n,k}}{z - i\lambda_{n,k}}$$

and so

$$f_\phi(z) = \sum_{k=1}^K \frac{a_{n,k}}{z - i\lambda_{n,k}} + \int_0^\infty [\phi(t) - \phi_n(t)] e^{-tz} dt.$$

Thus

$$|xf_\phi(x+iy) - M(e_{-y}\phi)| \leq \Big| \sum_{k=1}^{K} \frac{xa_{n,k}}{x+iy-i\lambda_{n,k}} - M(e_{-y}\phi) \Big| + \varepsilon.$$

The result now follows as in the proof of Lemma 7.3.2. □

PROOF OF THEOREM 7.3.7. The proof parallels that of Theorem 7.3.1. Suppose that f_ϕ is analytic across $(ia, ib) \subset i\mathbb{R}$. Then since $f(x+iy)$ is bounded for $y \in (a, b)$ it follows from Lemma 7.3.8 that $M(e_{-y}\phi) = 0$ for $y \in (a, b)$ and so $\Omega(\phi) \cap (a, b) = \emptyset$. From here it follows, from Remark 7.2.6, that ϕ can be approximated uniformly by ϕ_n of the form

$$\phi_n(t) = \sum_{k=1}^{K_n} a_{n,k} e^{it\lambda_{n,k}}, \quad \lambda_{n,k} \notin (a, b).$$

In fact, arrange things such that $|\phi(t) - \phi_n(t)| \leq 1/n$ for all t. As before

$$f_{\phi_n}(z) = \sum_{k=1}^{K} \frac{a_{n,k}}{z - i\lambda_{n,k}} := R_n(z)$$

(with a similar expression for $F_\phi(z)$). Also,

$$|f_\phi(z) - R_n(z)| \leq \int_0^\infty |\phi(t) - \phi_n(t)||e^{-tz}|dt \leq \frac{1}{n}\frac{1}{x}, \quad z = x+iy, \quad x > 0$$

$$|F_\phi(z) - R_n(z)| \leq \frac{1}{n}\frac{1}{|x|}, \quad z = x+iy, \quad x < 0.$$

The proof is finished by applying an analogous argument as at the end of the proof of Theorem 7.3.1. □

CHAPTER 8

Continuation by formal multiplication of series

In this chapter, we wish to introduce a new type of continuation inspired by the proof of the Douglas-Shapiro-Shields theorem (Theorem 6.3.4) which characterized the non-cyclic vectors for the backward shift S^* on H^2: $f \in H^2$ is non-cyclic for S^* if and only if f has a pseudocontinuation across $\mathbb{T}$ to a function $C_f \in \mathfrak{M}(\mathbb{D}_e)$ of bounded type. To motivate the definition of our new continuation, we would like to outline a proof of Theorem 6.3.4. Indeed, suppose that

$$f = \sum_{j=0}^{\infty} a_j z^j \in H^2$$

is non-cyclic for S^*. Then there is a non-trivial

$$g = \sum_{j=0}^{\infty} b_j z^j \in H^2$$

such that

$$\langle S^{*n} f, g \rangle = \sum_{j=0}^{\infty} a_{n+j} \overline{b_j} = 0 \text{ for all } n = 0, 1, \ldots$$

In fact, one can even take g to be a bounded function [1]. If one defines

$$G(z) = \sum_{j=0}^{\infty} \frac{\overline{b_j}}{z^j} \quad \text{and} \quad F(z) = \sum_{j=1}^{\infty} \frac{A_j}{z^j}, \quad z \in \mathbb{D}_e,$$

where

$$A_j = \sum_{k=0}^{\infty} a_k \overline{b_{k+j}},$$

then it is easy to see that F and G are analytic functions on $\mathbb{D}_e$ and $G \in H^\infty(\mathbb{D}_e)$. What takes slightly more work [51, p. 42] is to show that $F \in H^1(\mathbb{D}_e)$. So, using the fact that every Hardy space function is of bounded type, we conclude that F/G is of bounded type and so the radial limits of f (from the inside) and F/G (from the outside) exist almost everywhere. To prove these radial limit functions are the same, that is

$$f(e^{i\theta}) = \frac{F}{G}(e^{i\theta}) \text{ a.e.,}$$

we will show that $Gf = F$ almost everywhere, or what is the same, the Fourier coefficients of Gf are equal to those of F. The Fourier series for Gf is

$$\left\{ \overline{b_0} + \overline{b_1} e^{-i\theta} + \overline{b_2} e^{-2i\theta} + \cdots \right\} \left\{ a_0 + a_1 e^{i\theta} + a_2 e^{2i\theta} + \cdots \right\},$$

---

[1]From Chapter 6, § 6.3, the annihilator of the S^*-invariant subspace generated by f is a non-zero forward shift invariant subspace of H^2, which, by Beurling's theorem, is of the form IH^2 for some inner function I. Take $g = I$.

which, after gathering up like terms and making use of the identities

$$\sum_{j=0}^{\infty} a_{n+j}\overline{b_j} = \langle S^{*n} f, g \rangle = 0, \quad n = 0, 1, \ldots$$

$$\sum_{k=0}^{\infty} a_k \overline{b_{k+j}} = A_j, \quad j = 1, 2, \ldots,$$

is equal to

$$A_1 e^{-i\theta} + A_2 e^{-2i\theta} + \cdots.$$

But this last Fourier series is that of F. This proves that the boundary limits of f and F/G are equal almost everywhere and so F/G is of bounded type and is a pseudocontinuation (across $\mathbb{T}$) of f. It is not too difficult to show, using the F. and M. Riesz theorem (Theorem 2.2.4), that this process can be reversed yielding Theorem 6.3.4.

We would like to generalize this analysis beyond the H^2 setting by somehow associating the functions $f \in \mathfrak{H}(\mathbb{D})$ with $F/G \in \mathfrak{M}(\mathbb{D}_e)$ saying that, at least in terms of formal multiplication of Laurent series, Gf is equal to F. In this general a setting, the problem lies with the fact that general analytic (meromorphic) functions need not have boundary limits and, even if they did, they may not have L^1 boundary functions, making the Fourier series analysis used above impossible. Not being intimidated by such technicalities, we will define a continuation via 'formal Laurent series multiplication' and show that it is indeed meaningful, and, virtually the same analysis used above, but without mentioning Fourier series, can be used to discuss the cyclic vectors for the backward shift operator on other spaces of analytic functions such as the Dirichlet type spaces

$$D_\alpha = \Big\{\, f = \sum_{n=0}^{\infty} a_n z^n \in \mathfrak{H}(\mathbb{D}) : \sum_{n=0}^{\infty} (1+n)^\alpha |a_n|^2 < \infty \,\Big\}$$

as well as

$$\ell_A^p = \Big\{\, f = \sum_{n=0}^{\infty} a_n z^n \in \mathfrak{H}(\mathbb{D}) : \sum_{n=0}^{\infty} |a_n|^p < \infty \,\Big\}.$$

See Chapter 2 for a review of these spaces.

8.1. Definition

Suppose that $f \in \mathfrak{H}(\mathbb{D})$ and $G \in \mathfrak{H}(\mathbb{D}_e)$ have power series representations

$$f(z) = \sum_{j=0}^{\infty} a_j z^j, \quad z \in \mathbb{D}, \qquad G(z) = \sum_{j=0}^{\infty} \frac{B_j}{z^j}, \quad z \in \mathbb{D}_e.$$

Suppose also that for each $j = 0, 1, \ldots$ the two series

(8.1.1)
$$A_j := \sum_{k=0}^{\infty} a_k B_{j+k}$$

(8.1.2)
$$\widetilde{A_j} := \sum_{k=0}^{\infty} a_{k+j} B_k$$

converge absolutely. These two conditions allow us to formally multiply out the product

$$\left\{\, a_0 + a_1 z + a_2 z^2 + \cdots \right\}\left\{ B_0 + \frac{B_1}{z} + \frac{B_2}{z^2} + \cdots \right\}$$

and gather up like terms in powers of z and $1/z$ (without worrying about rearrangement of the terms) to form the formal Laurent series

(8.1.3)
$$f\#G := \sum_{j=0}^{\infty} \widetilde{A_j} z^j + \sum_{j=1}^{\infty} \frac{A_j}{z^j}.$$

REMARK 8.1.4. It is important to remember in eq.(8.1.3) that $f\#G$ is only a *formal* Laurent series and may not converge for any value of z.

DEFINITION 8.1.5. If $F = \sum_j C_j/z^j$ and $G = \sum_j B_j/z^j$ belong to $\mathfrak{H}(\mathbb{D}_e)$ and $f = \sum_j a_j z^j \in \mathfrak{H}(\mathbb{D})$, we say the meromorphic function F/G is a 'continuation by formal multiplication of series' of f if

(1) The series in eq.(8.1.1) and eq.(8.1.2) converge absolutely for each $j = 0, 1, 2, \ldots$
(2) The formal Laurent series $f\#G$ is equal to the Laurent series of F, *i.e.*,

$$\widetilde{A_j} = 0, \quad j = 0, 1, 2, \ldots \quad \text{and} \quad A_j = C_j, \quad j = 1, 2, \ldots$$

8.2. Compatibility with analytic continuation

Unfortunately, continuation by formal multiplication of series, henceforth denoted by (FM), is not in general unique (Theorem 8.5.1). However, with 'reasonable' assumptions on the coefficients $\{a_j\}$ and $\{B_j\}$, (FM) is at least compatible with analytic continuation. This is contained in the following 'Abelian' theorem.

THEOREM 8.2.1. *Let F/G be a (FM) continuation of f (as in Definition 8.1.5). Furthermore, suppose there exists a function $M : [0, 1) \to [0, \infty)$ such that*

$$\int_0^1 \log^+ \log^+ M(r)\, dr < \infty$$

and for all $z \in \mathbb{D}$,

$$\sum_{j=0}^{\infty} |B_j||z|^j \le M(|z|)$$

$$\sum_{t=0}^{\infty} \left\{ \sum_{j=0}^{\infty} |a_j||B_{j+t}| \right\} |z|^t \le M(|z|)$$

$$\sum_{t=0}^{\infty} \left\{ \sum_{j=0}^{\infty} |a_{j+t}||B_j| \right\} |z|^t \le M(|z|).$$

If f has an analytic continuation to an open neighborhood U of $\zeta \in \mathbb{T}$, then $f = F/G$ on $U \cap \mathbb{D}_e$.

Although the hypotheses of the above theorem are somewhat technical, they are easily checked when applying this result to discuss properties of cyclic vectors for the backward shift operator on some well-known spaces of analytic functions (see Proposition 8.3.4).

The 'log log' hypothesis of the theorem suggests the use of Levinson's theorem somewhere down the road. This is indeed the case. We record the result here and refer the reader to Koosis' book [**90**, p. 376] for the proof. See Lemma 7.3.6 for a related result.

THEOREM 8.2.2 (Levinson). *Consider the region*

$$R = \{re^{i\theta} : \theta_1 < \theta < \theta_2, r_1 < r < r_2\},$$

where $r_1 < 1 < r_2$. Let $m : (r_1, r_1) \to (0, \infty]$ be Lebesgue measurable with

$$\int_{r_1}^{r_2} \log^+ \log^+ m(r) dr < \infty.$$

Then, there is a decreasing function $w : (0, \infty) \to (0, \infty)$, depending only on the function m, such that if f is analytic on R with

$$|f(z)| \leq m(|z|), \quad z \in R,$$

then

$$|f(z)| \leq w(\operatorname{dist}(z, \partial R)), \quad z \in R.$$

In terms of applications, a family $\mathcal{F} \subset \mathfrak{H}(R)$ satisfying the hypothesis of Levinson's theorem will be a normal family.

PROOF OF THEOREM 8.2.1. For each $N = 0, 1, \ldots$ define the function

$$G_N(z) := \sum_{n=0}^{N} \frac{B_n}{z^n}, \quad z \neq 0,$$

and the complex numbers

$$C_{N,s} := \sum_{j=0}^{N} a_{j+s} B_j, \quad s = 0, 1, 2, \ldots$$

$$C_{N,-t} := \sum_{j=0}^{N-t} a_j B_{j+t}, \quad t = 0, 1, \ldots, N.$$

Next, use the hypotheses of the theorem to show that the function H_N defined on $\mathbb{D}$ by

$$H_N(z) := \sum_{s=1}^{\infty} C_{N,s} z^s$$

is analytic on $\mathbb{D}$. Also, define the function

$$I_N(z) := \sum_{t=0}^{N} \frac{C_{N,-t}}{z^t}, \quad z \neq 0$$

and observe that a computation shows

(8.2.3) $$H_N(z) = f(z)G_N(z) - I_N(z), \quad z \in \mathbb{D}\backslash\{0\}.$$

If we assume that f has an analytic continuation (also denoted by f) to an open neighborhood U of $\zeta \in \mathbb{T}$, we use eq.(8.2.3) to conclude the same is true for the function H_N. Note that G_N and I_N are rational functions with a pole at $z = 0$. To finish the proof, we will produce a subsequence $\{N_k\}$ such that

$$(8.2.4) \qquad\qquad H_{N_k} \to 0$$

$$(8.2.5) \qquad\qquad I_{N_k} \to F$$

uniformly on compact subsets of $U \cap \mathbb{D}_e$. This, along with the fact that $G_{N_k} \to G$ uniformly on compact subsets of $U \cap \mathbb{D}_e$ and eq.(8.2.3), yield $f = F/G$ on $U \cap \mathbb{D}_e$.

In order to prove eq.(8.2.4) and eq.(8.2.5), we need to make some estimates. For $z \in \mathbb{D}$,

$$
\begin{aligned}
|H_N(z)| &\leq \sum_{s=1}^{\infty} |C_{N,s}||z|^s \\
&\leq \sum_{s=1}^{\infty} \left\{ \sum_{j=0}^{N} |a_{j+s}||B_j| \right\} |z|^s \\
&\leq \sum_{s=1}^{\infty} \left\{ \sum_{j=0}^{\infty} |a_{j+s}||B_j| \right\} |z|^s \\
&\leq M(|z|).
\end{aligned}
$$

For $z \in \mathbb{D}_e$,

$$
|G_N(z)| \leq \sum_{j=0}^{N} |B_j||z|^{-j} \leq \sum_{j=0}^{\infty} |B_j||z|^{-j} \leq M(1/|z|)
$$

$$(8.2.6) \qquad |I_N(z)| \leq \sum_{t=1}^{N} |C_{N,-t}||z|^{-t} \leq \sum_{t=0}^{\infty} \left\{ \sum_{j=0}^{\infty} |a_j||B_{j+t}| \right\} |z|^{-t} \leq M(1/|z|)$$

and so (from eq.(8.2.3)) for $z \in U \cap \mathbb{D}_e$,

$$
|H_N(z)| \leq |f(z)||G_N(z)| + |I_N(z)| \leq CM(1/|z|).
$$

In summary,

$$
|H_N(z)| \leq m(|z|), \quad z \in U, \quad N = 0, 1, 2, \ldots
$$

where

$$
m(r) := \begin{cases} M(r) & r < 1, \\ \infty & r = 1, \\ M(1/r) & r > 1. \end{cases}
$$

The function m satisfies the hypothesis of Levinson's theorem (Theorem 8.2.2) and so $\{H_N\}$ forms a normal family on U. From eq.(8.2.6), $\{I_N\}$ also forms a normal family on $\mathbb{D}_e$. From here, one can find a subsequence $\{N_k\}$ so that $\{H_{N_k}\}$ and $\{I_{N_k}\}$ converge uniformly on compact subsets of U (respectively $\mathbb{D}_e$).

Since $\{I_{N_k}\}$ converges uniformly on compact subsets of $\mathbb{D}_e$ to some $I \in \mathfrak{H}(\mathbb{D}_e)$, which we denote by

$$
I := \sum_{t=1}^{\infty} \frac{C_{-t}}{z^t},
$$

then (by Cauchy's theorem) for each t, the numbers $C_{N_k,-t}$ converge to C_{-t} as $N_k \to \infty$. Recalling that

$$A_t = \sum_{j=0}^{\infty} a_j B_{j+t}, \quad t = 1, 2, \ldots$$

are the Taylor coefficients (about $z = \infty$) of F, observe that

$$C_{N_k,-t} - A_t = \sum_{j=0}^{N_k-t} a_j B_{j+t} - \sum_{j=0}^{\infty} a_j B_{j+t} = \sum_{j=N_k-t+1}^{\infty} a_j B_{j+t}$$

which, by the assumption in eq.(8.1.1), is the 'tail end' of an absolutely convergent series and so $C_{N_k,-t} \to A_t$. Thus

$$I_{N_k}(z) \to \sum_{t=1}^{\infty} \frac{A_t}{z^t} = F(z)$$

uniformly on compact subsets of $\mathbb{D}_e$, which proves eq.(8.2.5).

To prove eq.(8.2.4), recall that H_{N_k} converges uniformly on compact subsets of U and so it suffices to prove $H_{N_k}(z) \to 0$ for each $z \in U \cap \mathbb{D}$. To this end, notice that for each $z \in U \cap \mathbb{D}_e$,

$$|H_{N_k}(z)| \le \sum_{s=1}^{\infty} |C_{N_k,s}||z|^s.$$

Furthermore, by the hypothesis $f \# G = F$, we have

$$0 = \widetilde{A_s} := \sum_{j=0}^{\infty} a_{j+s} B_j = C_{N_k,s} + \sum_{j=N_k+1}^{\infty} a_{j+s} B_j, \quad s = 0, 1, 2, \ldots,$$

and so

$$|H_{N_k}(z)| \le \sum_{s=1}^{\infty} |C_{N_k,s}||z|^s \le \sum_{s=1}^{\infty} \left\{ \sum_{j=N_k+1}^{\infty} |a_{j+s}||B_j| \right\} |z|^s$$

which converges to zero by our hypothesis and the dominated convergence theorem. This proves eq.(8.2.4) and the proof of the theorem is now complete. $\square$

8.3. Cyclic vectors for backward shifts

The (FM) continuation from the previous section can be used to investigate the cyclic vectors for the backward shift operator on a variety of spaces of analytic functions on $\mathbb{D}$ such as the 'Dirichlet' type spaces D_α (see Chapter 2).

It is routine to check that $Bf \in D_\alpha$ whenever $f \in D_\alpha$, where B is the 'backward shift operator'

$$(Bf)(z) = \frac{f(z) - f(0)}{z},$$

and so B is a continuous linear operator on D_α. Just as we did for the classical Hardy, Bergman, and Dirichlet spaces in Chapter 6 (§ 6.3 and § 6.6), we investigate the 'cyclic vectors' for B, that is, those $f \in D_\alpha$ for which

$$\bigvee \{ B^n f : n = 0, 1, \ldots \} = D_\alpha.$$

As we have seen previously, especially for the classical Dirichlet space D_1, a complete characterization of the cyclic vectors for D_α is very much an open question. However, we will prove shortly, using our recent compatibility result (Theorem 8.2.1), that if $f \in D_\alpha$ has an isolated winding point at $\zeta_0 \in \mathbb{T}$ (see Example 6.2.3), then f is cyclic. We will also explore the question of whether or not the non-cyclic vectors form a linear manifold.

To proceed, we reformulate the condition for cyclicity by using the 'Cauchy duality' (see Chapter 2) which identifies the dual of D_α with $D_{-\alpha}$ via

$$\langle f, g \rangle = \sum_{n=0}^{\infty} a_n \overline{b_n},$$

where $\{a_n\}$ are the Taylor coefficients of $f \in D_\alpha$ and $\{b_n\}$ are those of $g \in D_{-\alpha}$. By the Hahn-Banach theorem, $f = \sum_j a_j z^j \in D_\alpha$ is *not* cyclic if and only if there is a non-trivial $g = \sum_j b_j z^j \in D_{-\alpha}$ such that

(8.3.1) $\langle B^n f, g \rangle = \sum_{j=0}^{\infty} a_{n+j} \overline{b_j} = 0$ for all $n = 0, 1, 2, \dots$

If

$$D_\alpha(\mathbb{D}_e) := \{G(z) = g(1/z) : g \in D_\alpha\},$$

one notes that for the annihilating $g = \sum_j b_j z^j$ as above, the function

$$G(z) := \sum_{j=0}^{\infty} \frac{\overline{b_j}}{z^j}$$

belongs to $D_{-\alpha}(\mathbb{D}_e)$ and, using eq.(8.3.1), the formal product $f \# G$ becomes

(8.3.2) $f \# G = \sum_{j=1}^{\infty} \frac{A_j}{z^j}, \quad A_j = \sum_{k=0}^{\infty} a_k \overline{b_{j+k}}.$

Observe that

$$
\begin{aligned}
|A_j| &\leq \sum_k |a_k||b_{j+k}| \\
&= \sum_k |a_k|(1+k)^{\alpha/2}|b_{j+k}|(1+j+k)^{-\alpha/2} \left[\frac{1+j+k}{1+k} \right]^{\alpha/2} \\
&\leq \sup_k \left[\frac{1+j+k}{1+k} \right]^{\alpha/2} \sqrt{\sum_k |a_k|^2(1+k)^\alpha} \sqrt{\sum_k |b_{j+k}|^2(1+j+k)^{-\alpha}}
\end{aligned}
$$

which is $O(j^{\alpha/2})$ when $\alpha > 0$ and $O(1)$ when $\alpha \leq 0$. In either case,

(8.3.3) $|A_j| = O(j^N)$

for some $N \in \mathbb{N}$ and so the formal Laurent series in eq.(8.3.2) converges uniformly on compact subsets of $\mathbb{D}_e$ to a function F belonging to $\mathfrak{H}(\mathbb{D}_e)$.

We have thus shown that if $f \in D_\alpha$ is not cyclic, then there is a non-trivial $G \in D_{-\alpha}(\mathbb{D}_e)$ and an $F \in \mathfrak{H}(\mathbb{D}_e)$ with $F(\infty) = 0$ such that $f \# G = F$. One checks that the above argument can be reversed which leads us to the following proposition.

PROPOSITION 8.3.4. *A vector $f \in D_\alpha$ is not cyclic if and only if there is a non-trivial $G \in D_{-\alpha}(\mathbb{D}_e)$ and an $F \in \mathfrak{H}(\mathbb{D}_e)$ with $F(\infty) = 0$ such that $f \# G = F$.*

REMARK 8.3.5.   (1) In the Hardy space case ($\alpha = 0$), the function F can be chosen to belong to $H^1(\mathbb{D}_e)$ [**51**, proof of Thm. 2.2.1] and G can be taken to be bounded.

(2) It is interesting to remark here that this theorem not only solves the cyclicity problem, although in a rather weak sort of way, but also shows that there are functions which have a (FM) continuation but not a pseudocontinuation. Consider the function f on the disk defined by the Borel series in eq.(6.6.6). This function was constructed to be a non-cyclic vector for the classical Dirichlet space D_1 and hence, by the above proposition, has a (FM) continuation. However, this function was also constructed to not have a pseudocontinuation across any arc of $\mathbb{T}$.

(3) Although Proposition 8.3.4 is a necessary and sufficient condition for cyclicity, it is almost impossible to apply to particular examples. However, if we combine this result with Theorem 8.2.1, we do obtain some new examples of cyclic vectors. Recall from Example 6.2.3, the compatibility of pseudocontinuation with analytic continuation, that in the H^2 setting, a function which has an isolated winding point at $\zeta_0 \in \mathbb{T}$ is a cyclic vector. Using Theorem 8.2.1, our compatibility result for (FM) continuation, one can obtain a similar result for D_α.

COROLLARY 8.3.6. *If $f \in D_\alpha$ has an isolated winding point at $\zeta_0 \in \mathbb{T}$, then f is cyclic.*

PROOF. One can quickly check that if $f = \sum_j a_j z^j \in D_\alpha$ and $G = \sum_j B_j/z_j \in D_{-\alpha}(\mathbb{D}_e)$, the 'majorization' function M in the hypothesis of Theorem 8.2.1 can be taken to be $M(r) = (1-r)^{-N}$ for some $N \in \mathbb{N}$. Indeed, by eq.(8.3.3), terms of the form

$$\sum_j |a_j||B_{j+t}| \quad \text{and} \quad \sum_j |a_{j+t}||B_j|$$

are $O(t^p)$ for some $p \geq 0$. Since

$$\sum_j |B_j|^2 (1+j)^{-\alpha} < \infty,$$

then $|B_j| = o(j^{\alpha/2})$. Such an M is clearly $\log^+ \log^+$ integrable on $[0,1)$.

Now supposing that $f \in D_\alpha$ has an isolated winding point at $\zeta \in \mathbb{T}$ but is not cyclic, we can apply Proposition 8.3.4 to conclude that f has a (FM) continuation F/G, where $F \in \mathfrak{H}(\mathbb{D}_e)$ and $G \in D_{-\alpha}(\mathbb{D}_e)$. However, due to the compatibility of (FM) continuation with analytic continuation (see Example 6.2.3), we have a contradiction. $\square$

Recall from Theorem 6.3.8 that if f_1 and f_2 are non-cyclic vectors for the backward shift on H^2, then the vector $f_1 + f_2$ is also non-cyclic. Indeed this is easy to show as follows: By the Douglas-Shapiro-Shields theorem (Theorem 6.3.4), f_1 has a pseudocontinuation of bounded type F_1/G_1 (where F_1 and G_1 are bounded analytic functions on $\mathbb{D}_e$) and similarly, f_2 also has a pseudocontinuation of bounded type F_2/G_2. The function $F_1/G_1 + F_2/G_2$ is also of bounded type and is the pseudocontinuation of $f_1 + f_2$.

When one tries the same analysis on non-cyclic vectors f_1 and f_2 on the Dirichlet type spaces D_α, some problems arise. For example, f_1 has a (FM) continuation F_1/G_1, where F_1 is analytic on $\mathbb{D}_e$, $F_1(\infty) = 0$, and $G_1 \in D_{-\alpha}(\mathbb{D}_e)$. Similarly, f_2 has a (FM) continuation F_2/G_2, where F_2 is analytic on $\mathbb{D}_e$, $F_2(\infty) = 0$, and $G_2 \in D_{-\alpha}(\mathbb{D}_e)$. At least formally, the function

$$\frac{F_1}{G_1} + \frac{F_2}{G_2} = \frac{F_1 G_2 + G_1 F_2}{G_1 G_2}$$

is a (FM) continuation of $f_1 + f_2$. However there is no reason *a priori* to believe that the product $G_1 G_2$ belongs to $D_{-\alpha}(\mathbb{D}_e)$ (as is needed to apply Proposition 8.3.4). In the Hardy space case, G_1 and G_2 were bounded functions and so $G_1 G_2$ is also bounded. Unfortunately, as we shall see in a moment, one can not always arrange $G_1 G_2$ to belong to $D_{-\alpha}(\mathbb{D}_e)$ and in fact, $f_1 + f_2$ may indeed be a cyclic vector. To make our presentation clear, we focus on two distinct representative cases: D_{-1} (the Bergman space) and D_1 (the Dirichlet space). We begin with a positive result.

THEOREM 8.3.7. *If f_1 and f_2 are non-cyclic vectors for the backward shift on the Bergman space D_{-1}, then $f_1 + f_2$ is also non-cyclic.*

PROOF. By a theorem of Richter and Sundberg [113, Thm. 3.2], every S-invariant subspace of the Dirichlet space is singly generated by a multiplier [2]. Thus we can assume that g_1 and g_2, where g_i is non-trivial and annihilates the B-invariant subspace generated by f_i, are multipliers of D_1. Thus the function

$$\mathcal{G} = G_1 G_2$$

also belongs to $D_1(\mathbb{D}_e)$, where $G_i(z) = g_i(1/z)$ as in the discussion preceding Proposition 8.3.4.

We need to show that $(f_1 + f_2)\#\mathcal{G}$ consists only of terms z^m with $m < 0$. It suffices to show this for $f_1\#\mathcal{G}$ and $f_2\#\mathcal{G}$ separately.

It is not too difficult to show that if $\mathcal{G}_n \to \mathcal{G}$ in Dirichlet norm of the exterior disk (*i.e.*, $\mathcal{G}_n(1/z)$ goes to $\mathcal{G}(1/z)$ in the usual Dirichlet norm), then for each $m \geq 0$, the coefficient of z^m in $f_1\#\mathcal{G}$ is the limit of the corresponding coefficient in $f_1\#\mathcal{G}_n$ as $n \to \infty$. Indeed, the z^m coefficient of $f_1\#\mathcal{G}$ is

$$a_m B_0 + a_{m+1} B_1 + \cdots,$$

where the a's are the Taylor coefficients of f_1 and the B's are those of $\mathcal{G}$. In a similar way, the z^m coefficient of $f_1\#\mathcal{G}_n$ is

$$a_m B_{n,0} + a_{m+1} B_{n,1} + \cdots.$$

The difference of these two numbers is bounded above, in absolute value, by

$$|a_m||B_0 - B_{n,0}| + |a_{m+1}||B_1 - B_{n,1}| + \cdots$$

$$\leq \Big(\sum_{j=0}^{\infty} \frac{|a_{m+j}|^2}{j+1} \Big)^{1/2} \Big(\sum_{j=0}^{\infty} (j+1)|B_j - B_{n,j}|^2 \Big)^{1/2}.$$

---

[2]An S-invariant subspace $\mathcal{K} \subset D_1$ is 'singly generated' if there is a $g \in D_1$ such that $\mathcal{K} = \bigvee\{S^n g : n = 0, 1, 2, \cdots\}$. A function $g \in D_1$ is a 'multiplier' of D_1 if $gD_1 \subset D_1$. It is routine to show that a multiplier must be a bounded function [53, Lemma 11]. The converse is not true.

From here one can see the above is bounded above by $C_{m,f}\|\mathcal{G} - \mathcal{G}_n\|$ which goes to zero as $n \to \infty$ since $\mathcal{G}_n \to \mathcal{G}$ in Dirichlet norm.

Now, suppose that $G_{n,2}$ are the 'Fejér means' of G_2, that is,

$$G_{n,2}(z) = \sigma_n(G_2)(z) = \frac{1}{n+1} \sum_{k=0}^{n} s_k(G_2)(z), \quad z \in \mathbb{D}_e,$$

where

$$s_k(G_2)(z) = \sum_{j=0}^{k} \frac{B_j}{z^j}$$

is the k-th partial sum of G_2. Since the z^m coefficient of $f_1 \# G_1$ is zero for $m \geq 0$, then, by a straightforward exercise, the z^m coefficient of $f \#(G_1 G_{n,2})$ is also zero for $m \geq 0$. In view of our comment above, we just need to show that

$$G_1 G_{n,2} \to G_1 G_2$$

in the Dirichlet norm of the exterior disk, that is

$$\int_{\mathbb{D}} |\frac{d}{dz}(g_1 g_{n,2} - g_1 g_2)|^2 dA(z) \to 0, \quad n \to \infty.$$

It suffices to show that the integrals

$$\int_{\mathbb{D}} |g_1|^2 |\frac{d}{dz}(g_{n,2} - g_2)|^2 dA(z) \quad \text{and} \quad \int_{\mathbb{D}} |g_{n,2} - g_2|^2 |g_1'|^2 dA(z)$$

go to zero as $n \to \infty$. The second integral goes to zero (as $n \to \infty$) since g_2 is a bounded function (since it is a multiplier) and hence the Fejér means are uniformly bounded in sup-norm [**77**, pp. 16 - 19]. Now use the dominated convergence theorem along with the fact that $g_{n,2}$ converge pointwise to g_2 in $\mathbb{D}$. For the first integral, use the fact that g_1 is a bounded function as well as the fact that $s_k(g_2) \to g_2$ in the Dirichlet norm (i.e., $s_k(g_2)' \to g_2'$ in the $L^2(dA)$ norm) and hence the Fejér means $g_{n,2}$ also converge to g_2 in Dirichlet norm. $\qquad\square$

The negative result is the following.

THEOREM 8.3.8. *There are two non-cyclic vectors f_1 and f_2 for the backward shift on the Dirichlet space D_1 whose sum $f_1 + f_2$ is cyclic.*

REMARK 8.3.9. This result was originally discovered by Abakumov [**1**, p. 77] (see also [**2**]) using a gap series argument. The argument to be presented here uses zero sequences of Bergman spaces and the coherence results of Gončar presented earlier (Theorem 5.2.3). Our argument also has several generalizations.

Before proceeding to the proof, we need to bring in a few facts from the theory of Bergman spaces. The general idea to prove Theorem 8.3.8 will be to define f_1 and f_2 by Borel series

$$f_1 = \sum_{n=1}^{\infty} \frac{C_n}{1 - \overline{a}_n z}, \quad f_2 = \sum_{n=1}^{\infty} \frac{C_n}{1 - \overline{b}_n z},$$

where $\{a_n\}, \{b_n\} \subset \mathbb{D}$ are the zeros of some non-trivial functions in the Bergman space D_{-1} and the C_n's are sufficiently small so that f_1 and f_2 belong to the Dirichlet space D_1. Note that if g is a non-trivial function in D_{-1} with $g(\{a_n\}) = 0$

(certainly possible since $\{a_n\}$ are the zeros of some non-trivial Bergman function), then

$$\langle B^N f_1, g \rangle = \sum_{n=1}^{\infty} C_n \overline{a_n}^N \overline{g(a_n)} = 0, \quad N = 0, 1, 2, \ldots$$

and so f_1 is a non-cyclic vector. Similarly, f_2 is non-cyclic. The sum $f_1 + f_2$ will also be a Borel series and we claim that the coefficients $\{C_n\}$ as well as the zero sequences $\{a_n\}$ and $\{b_n\}$ can be chosen so that $f_1 + f_2$ will be our desired cyclic vector.

To this end, we recall a result of C. Horowitz [**79**, Thm. 2] which says it is possible to have two disjoint zero sequences $\{a_n\}$ and $\{b_n\}$ of the Bergman space whose union $\{a_n\} \cup \{b_n\}$ is not a zero set [3]. Let the union of these two sequences be denoted by $\{z_n\}$. Theorem 8.3.8 will follow, by choosing appropriate (rapidly decreasing) C_n's, from the following more general result.

THEOREM 8.3.10. *Let $\alpha \in \mathbb{R}$ and let $\{z_n\}$ be a sequence of distinct points in the disk with $|z_n| \to 1$ and such that $\{z_n\}$ is not a zero sequence of any non-trivial function from $D_{-\alpha}$. If $\{C_n\}$ is a sequence of non-zero complex numbers such that the Borel series*

$$f := \sum_{n=1}^{\infty} \frac{C_n}{1 - \overline{z_n} z}$$

belongs to D_α and such that

$$\varlimsup_{n \to \infty} \sqrt[n]{|C_n|\, h_\alpha(|z_n|)} < 1,$$

then f is a cyclic vector for the backward shift on D_α.

REMARK 8.3.11. Recall from Chapter 2 that

$$h_\alpha(t) := \begin{cases} (1-t)^{-(1-\alpha)/2} & \alpha < 1, \\ \sqrt{\log \frac{1}{1-t}} & \alpha = 1, \\ 1 & \alpha > 1 \end{cases}$$

Moreover, for any $g \in D_\alpha$

(8.3.12) $|g(z)| = A_\alpha \|g\|_{D_\alpha} h_\alpha(|z|), \quad z \in \mathbb{D}.$

PROOF OF THEOREM 8.3.10. It suffices to show that the only $g \in D_{-\alpha}$ such that

$$\langle B^N f, g \rangle = 0 \text{ for all } N = 0, 1, 2, \ldots$$

is the zero function. First note that

$$B^N \frac{1}{1 - \overline{z_n} z} = \overline{z_n}^N \frac{1}{1 - \overline{z_n} z}$$

and moreover,

$$\left\langle \frac{1}{1 - \overline{z_n} z}, g \right\rangle = \overline{g(z_n)}.$$

---

[3]In fact, one can go further and arrange $\{a_n\} \cup \{b_n\}$ to be a 'sampling sequence' (see [**70**, **123**, **124**]) for the Bergman space.

Thus

$$0 = \langle B^N f, g \rangle = \sum_{n=1}^{\infty} C_n \overline{z_n}^N \overline{g(z_n)}, \quad N = 0, 1, 2, \dots$$

From this follows

$$z^N \sum_{n=1}^{\infty} C_n \overline{z_n}^N \overline{g(z_n)} = 0, \quad N = 0, 1, 2, \dots, \quad |z| < 1.$$

To simplify notation, let $\lambda_n = C_n \overline{g(z_n)}$ and note, from eq.(8.3.12) and our hypothesis, that

$$\varlimsup_{n \to \infty} \sqrt[n]{|\lambda_n|} < 1.$$

Summing the above series on N and reversing the order of summation we get

$$\sum_{N=0}^{\infty} z^N \sum_{n=1}^{\infty} \lambda_n \overline{z_n}^N = \sum_{n=1}^{\infty} \frac{\lambda_n}{1 - \overline{z_n} z} = 0, \quad |z| < 1.$$

By the condition $\varlimsup_{n \to \infty} \sqrt[n]{|\lambda_n|} < 1$, the above Borel series has a Gončar continuation given by

$$\sum_{n=1}^{\infty} \frac{\lambda_n}{1 - \overline{z_n} z}, \quad |z| > 1.$$

By the compatibility of Gončar continuation with analytic continuation (Theorem 5.2.3), this function must be the zero function and so, by Cauchy's residue theorem,

$$\lambda_n = C_n \overline{g(z_n)} = 0 \quad \text{for all } n = 0, 1, 2, \dots$$

But since $\{z_n\}$ is not the zero sequence of any non-trivial $D_{-\alpha}$ function, we must conclude that $g \equiv 0$. $\qquad \square$

REMARK 8.3.13.   (1) The above analysis (Proposition 8.3.4 and Corollary 8.3.6) for the characterization of the cyclic vectors for D_α via (FM) continuations can be carried out in much the same way for the other spaces of analytic functions such as ℓ_A^p $(1 < p < \infty)$. It is well-known (and easy to prove) that the dual of ℓ_A^p can be identified with ℓ_A^q, where q is the conjugate index to p, via the pairing

$$\langle f, g \rangle = \sum_{n=0}^{\infty} a_n \overline{b_n},$$

where $\{a_n\}$ are the Taylor coefficients of $f \in \ell_A^p$ and $\{b_n\}$ are those of $g \in \ell_A^q$. The reader can check that Proposition 8.3.4 transforms in the ℓ_A^p setting to the following: $f \in \ell_A^p$ is non-cyclic if and only if there is a non-trivial $G \in \ell_A^q(\mathbb{D}_e)$ and an $F \in \mathfrak{H}(\mathbb{D}_e)$ with $F(\infty) = 0$ such that $f \# G = F$. Also note that Corollary 8.3.6 also applies in the ℓ_A^p setting (It is easy to check that the hypothesis of Theorem 8.2.1 are satisfied) and so functions in ℓ_A^p which have an isolated winding point $\zeta_0 \in \mathbb{T}$ are cyclic vectors.

(2) As mentioned earlier (see Remark 8.3.9), Theorem 8.3.8 is not the first instance where the set of non-cyclic vectors in some Banach space of analytic functions do not form a linear manifold. Nor is it the only one. For the spaces ℓ_A^p $(1 \le p < 2)$, Abakumov [2, Prop. 4.1] constructed a

non-cyclic vector f and a cyclic vector g such that $f + \lambda g$ is non-cyclic for every complex number λ.

The Borel series trick in Theorem 8.3.8 can also be used to prove some other interesting results.

THEOREM 8.3.14. *For each $\alpha \in \mathbb{R}$ there is a vector $f \in D_\alpha$ such that f is cyclic for B^2 on D_α.*

PROOF. Let $\{a_n\} \subset (0,1)$ with $a_n \to 1/2$ as $n \to \infty$. Since the vectors $(1 - a_n z)^{-1}$ have uniformly bounded D_α norms, the vector

$$f := \sum_{n=1}^{\infty} \frac{2^{-n}}{1 - a_n z}$$

belongs to D_α. Suppose that $g \in D_{-\alpha}$ satisfies

$$\langle B^{2N} f, g \rangle = 0 \text{ for all } N = 0, 1, 2, \ldots$$

Then, as in the proof of Theorem 8.3.10, we get that

$$\sum_{n=1}^{\infty} \frac{2^{-n} \overline{g(a_n)}}{1 - a_n^2 z} = 0 \text{ for all } |z| < 1.$$

As before, using the $1/2$-hyperconvergence of the Borel series (notice the a_n^2's are distinct!), we conclude that $g(\{a_n\}) = 0$, forcing the zeros of g to accumulate in $\mathbb{D}$. Thus g is identically zero and so f is cyclic.   □

The characterization of the cyclic vectors for B^2 is still very much an open problem, even in the classical H^2 case. Certainly any H^2 function which belongs to $PCBT$, and hence non-cyclic for B, is non-cyclic for B^2. The converse is false. For example, $f = \log(1 - z^2)$ is cyclic for B on H^2, due to its isolated winding points (Example 6.2.3). However, since anything in the B^2-invariant subspace generated f has only even powers in its Taylor series expansion, the function f is non-cyclic for B^2. By a similar argument, the gap series

$$\sum_{n=0}^{\infty} 2^{-n} z^{2^n}$$

is cyclic for B (Theorem 6.9.7) but non-cyclic for B^2. The above result produces cyclic vectors for B^2. Unfortunately, a complete characterization of them is unknown.

QUESTION 8.3.15. What are the cyclic vectors for B^2 on D_α?

A trivial modification of the proof of Theorem 8.3.14 shows that the f produced there is not only cyclic for B^2 but cyclic for B^N for every $N = 1, 2, \ldots$. This 'universal cyclic vector' problem has been investigated by others. For example, in the H^2 setting, notice that $B^N = T_{\bar{z}^N}$, a co-analytic Toeplitz operator on H^2. A result of Wogen [145] produces a vector $f \in H^2$ which is cyclic for *every* co-analytic Toeplitz operator $T_{\bar{\phi}}$, where $\phi \in H^\infty$ and not a constant function.

This 'universal cyclicity' is true in the most general setting as was shown in the following result of Bourdon and J. Shapiro [30]: Suppose $\mathcal{H}$ is a Hilbert space of analytic functions on a planar region Ω with the property that for each $a \in \Omega$, the evaluation functional $f \to f(a)$ is continuous on $\mathcal{H}$. Then there is a vector $f \in \mathcal{H}$ such that f is cyclic for M_ϕ^* for every non-constant multiplier[4] ϕ of $\mathcal{H}$. Here M_ϕ is the operator 'multiplication by ϕ' on $\mathcal{H}$. To give the reader a taste of what goes on here, without getting into any of the technical details, we prove the following weaker result.

PROPOSITION 8.3.16. *Let $\mathcal{H}$ be a Hilbert space of analytic functions on a planar domain Ω such that for each $a \in \Omega$ the evaluation functional $f \to f(a)$ is continuous on $\mathcal{H}$. Then, for a given non-constant multiplier ϕ of $\mathcal{H}$, there is a vector $f \in \mathcal{H}$ such that f is cyclic for M_ϕ^*.*

PROOF. Without loss of generality, assume that $0 \in \Omega$. Let ϕ be a given non-constant multiplier of $\mathcal{H}$ and choose a bounded sequence $\{a_n\} \subset \Omega$ which converges to a point of Ω and such that ϕ is one-to-one on $\{a_n\}$ and $\phi(a_n) \not\to 0$ as $n \to \infty$. Since, for each $a \in \Omega$, the evaluation functional $g \to g(a)$ is continuous on $\mathcal{H}$, we can apply the Riesz representation theorem to produce a unique $k_a \in \mathcal{H}$ such that $\langle g, k_a \rangle = g(a)$ for all $g \in \mathcal{H}$. The set of $\{k_a : a \in \Omega\}$ are called the 'reproducing kernels' of $\mathcal{H}$. One can easily argue [30, Lemma 1.2] that since the a_n's accumulate in Ω, the sequence $\{k_{a_n}\}$ is norm bounded. Thus the function

$$f := \sum_{n=1}^{\infty} 2^{-n} k_{a_n}$$

belongs to $\mathcal{H}$.

To show that f is cyclic for M_ϕ^*, we suppose that $g \in \mathcal{H}$ with

$$\langle M_\phi^{*N} f, g \rangle = 0 \quad \text{for all } N = 0, 1, 2, \dots$$

Then

$$0 = \langle M_\phi^{*N} f, g \rangle = \langle f, \phi^N g \rangle = \sum_{n=1}^{\infty} 2^{-n} \overline{\phi}(a_n)^N \overline{g}(a_n) \quad \text{for all } N = 0, 1, \dots$$

Since $\{\phi(a_n)\}$ is a bounded sequence (since multipliers are bounded [30, Prop. 1.2]), then for $z \in \mathbb{C}$ such that $|z| \sup\{|\phi(a_n)| : n \in \mathbb{N}\} < 1$ we can write (as in the proof of Theorem 8.3.10)

$$0 = \sum_{N=0}^{\infty} z^N \sum_{n=1}^{\infty} 2^{-n} \overline{\phi}(a_n)^N \overline{g}(a_n) = \sum_{n=1}^{\infty} \frac{2^{-n} \overline{g}(a_n)}{1 - \overline{\phi}(a_n) z}.$$

Since the sequence $\{\overline{g}(a_n)\}$ is a bounded sequence, the above Borel series $1/2$-hyperconverges (note that the $\overline{\phi}(a_n)$'s are distinct!) on compact subsets of

$$\mathbb{C} \backslash \{1/\overline{\phi}(a_n)\}^{-}.$$

Moreover, it converges to zero on the connected component of $\mathbb{C} \backslash \{1/\overline{\phi}(a_n)\}^{-}$ which contains the origin. As in the proof of Theorem 8.3.14, using the $1/2$-hyperconvergence of the above Borel series, we conclude that $g \equiv 0$ and so f is cyclic for M_ϕ^*. □

---

[4] A function ϕ is a 'multiplier' for $\mathcal{H}$ if $\phi f \in \mathcal{H}$ for every $f \in \mathcal{H}$. An application of the closed graph theorem shows that if ϕ is a multiplier of $\mathcal{H}$, then the operator $M_\phi : \mathcal{H} \to \mathcal{H}$ defined by $M_\phi f = \phi f$ is continuous on $\mathcal{H}$.

REMARK 8.3.17. What makes the f in the proof of the previous result not quite cyclic for every M_ϕ^* is the dependence of the choice of $\{a_n\}$ on ϕ. In order to get the hyperconvergence, ϕ needs to be one-to-one on $\{a_n\}$. Bourdon and Shapiro [**30**] do quite a bit more work to show that the sequence $\{a_n\}$ can be chosen independently of the multiplier ϕ.

We have already seen that for certain spaces of analytic functions, the classical Dirichlet space D_1 for example, the non-cyclic vectors for the backward shift do *not* form a linear manifold. We leave it to the reader to use the Borel series and reproducing kernel techniques used above, as well as the proof of Theorem 8.3.10, to prove the following theorem.

THEOREM 8.3.18. *Let $\mathcal{H}$ be a Hilbert space of analytic functions on a planar domain Ω such that for each $a \in \Omega$ the evaluation functional $f \to f(a)$ is continuous on $\mathcal{H}$. Moreover, suppose that $\mathcal{H}$ also has the property that there exist two disjoint zero sequences for $\mathcal{H}$ whose union is not a zero sequence for $\mathcal{H}$. Then for every non-constant multiplier ϕ of $\mathcal{H}$, there are two non-cyclic vectors for M_ϕ^* whose sum is cyclic.*

On can prove slightly more: there are two vectors which are non-cyclic for every operator in the set

$$U = \{M_\phi^* : \phi \text{ is a non-constant univalent multiplier of } \mathcal{H}\}$$

but whose sum is cyclic for every operator in U.

8.4. Spectral properties

We begin with some general remarks about the spectrum of the backward shift operator B on D_α. Our discussion here will be quite general and apply to a large class of Banach spaces of analytic functions beyond D_α (for example ℓ_A^p) which satisfy certain, unfortunately, rather cumbersome technical conditions. In order not to lose the reader in all the smoke, our discussion will begin in the D_α setting. At the very end, we will discuss the general Banach space situation.

First notice that

$$\frac{1}{z - \lambda} \in D_\alpha, \quad \text{for all } \lambda \in \mathbb{D}_e,$$

and

(8.4.1) $$B \frac{1}{z - \lambda} = \frac{1}{\lambda} \frac{1}{z - \lambda}, \quad B1 = 0.$$

Thus $\mathbb{D} \subset \sigma_p(B)$ (the point spectrum of B). It is routine to check that for $f \in D_\alpha$ and $\lambda \in \mathbb{D}$,

$$R_\lambda(f) := \frac{zf - \lambda f(\lambda)}{z - \lambda} \in D_\alpha.$$

Moreover, an application of the closed graph theorem shows that R_λ is a continuous operator on D_α and a computation shows that $(I - \lambda B)^{-1} = R_\lambda$. Thus

$$\sigma(B) = \mathbb{D}^-.$$

Here $\sigma(B)$ is the spectrum of the operator B.

Our focus in this section will be on the spectral properties of $B|\mathcal{M}$, where $\mathcal{M}$ is a B-invariant subspace of D_α. Probably the first to explore the spectral properties of $B|\mathcal{M}$ was Moeller [**100**] who showed, for a general B-invariant subspace $(\phi H^2)^\perp$ of H^2, where ϕ is an inner function, that

$$\sigma(B|(\phi H^2)^\perp) = \{\overline{\lambda} : \lambda \in \sigma(\phi)\},$$

where

$$\sigma(\phi) := \Big\{ \lambda \in \mathbb{D}^- : \varliminf_{z\to\lambda} |\phi(z)| = 0 \Big\}$$

is the 'liminf zero set' of ϕ (often called the 'spectrum' of ϕ). If $\phi = bs_\mu$, where b is the Blaschke factor of ϕ and s_μ is the singular inner factor, then $\sigma(\phi)$ is the closure of the zeros of b together with the support of the singular measure μ.

From Theorem 6.3.9, we know that every function in $(\phi H^2)^\perp$ has a pseudo-continuation across $\mathbb{T}$ to a function on $\mathbb{D}_e$ of bounded type. Using the fact that an inner function ϕ has an analytic continuation to

$$\mathbb{C}_\infty \backslash \{1/\overline{z} : z \in \sigma(\phi)\}$$

one can show [**51**, Cor. 3.1.10], using a Morera's theorem argument, that *every* function in $(\phi H^2)^\perp$ has an analytic continuation to the same set. The point here is that the spectral properties of the operator $B|(\phi H^2)^\perp$ are closely related to the 'continuation' properties of *all* the functions in our invariant subspace. We also point out that Aleksandrov [**4**] (see also [**35**, p. 185]) was able to describe the parts of the spectrum of $B|\mathcal{M}$ where $\mathcal{M}$ is a B-invariant subspace of H^p $(0 < p < 1)$. The description is a bit more complicated and is beyond what we want to do here.

Although, for a general B-invariant subspace $\mathcal{M}$ of D_α, we cannot prove a Moeller type theorem precisely identifying the spectrum of $B|\mathcal{M}$, we can still make a connection between the spectrum of $B|\mathcal{M}$ and the continuation properties of functions in $\mathcal{M}$. To avoid cumbersome notation, we make the following definition.

DEFINITION 8.4.2. Let $\alpha \in \mathbb{R}$ be fixed and $\mathcal{M}$ be a non-trivial B-invariant subspace of D_α. Define

$$T := B|\mathcal{M}.$$

From the spectral radius formula [**37**, p. 202] and the identity $\sigma(B) = \mathbb{D}^-$, follows

$$\sigma(T) \subset \mathbb{D}^-.$$

What requires slightly more work is the following.

PROPOSITION 8.4.3.
$$\sigma_{ap}(T) \cap \mathbb{D} = \sigma_p(T) \cap \mathbb{D} = \Big\{ a \in \mathbb{D} : \frac{1}{1-az} \in \mathcal{M} \Big\}.$$

REMARK 8.4.4. $\sigma_{ap}(T)$, the 'approximate point spectrum' [**37**, p. 213], is the set of points $w \in \mathbb{C}$ for which there is a sequence of unit vectors $\{f_n\}$ in $\mathcal{M}$ such that

$$\|(wI - T)f_n\| \to 0 \text{ as } n \to \infty.$$

A standard fact from operator theory is

$$w \notin \sigma_{ap}(T) \Leftrightarrow \ker(wI - T) = (0) \text{ and the range of } (wI - T) \text{ is closed.}$$

PROOF OF PROPOSITION 8.4.3. From eq.(8.4.1) follows

$$\sigma_p(T) \cap \mathbb{D} = \left\{ a \in \mathbb{D} : \frac{1}{1 - az} \in \mathcal{M} \right\}.$$

So we just need to verify $\sigma_{ap}(T) \cap \mathbb{D} = \sigma_p(T) \cap \mathbb{D}$. Clearly $\sigma_p(T) \subset \sigma_{ap}(T)$. For the other direction, suppose that $\lambda \in \sigma_{ap}(T) \cap \mathbb{D}$. Then, by Remark 8.4.4, there is a sequence $\{f_n\} \subset \mathcal{M}$ with

$$\|f_n\| = 1 \quad \text{and} \quad \|(T - \lambda I) f_n\| \to 0.$$

Since the inclusion mapping $i : D_\alpha \to \mathfrak{H}(\mathbb{D})$ is continuous, $\{f_n(0)\}$ is a bounded sequence of complex numbers and so there is a convergent subsequence (which we also call $\{f_n(0)\}$) converging to some $c \in \mathbb{C}$. Since

$$\left\| \frac{f_n - f_n(0)}{z} - \lambda f_n \right\| \to 0,$$

and the shift operator S ('multiplication by z') is continuous on D_α,

$$\|(1 - \lambda z) f_n - c\| \to 0.$$

Since $\sigma(S) = \mathbb{D}^-$, we conclude

$$f_n \to \frac{c}{1 - \lambda z}$$

and since $\|f_n\| = 1$, $c \neq 0$. Thus by eq.(8.4.1), $\lambda \in \sigma_p(T)$. $\square$

Since the polynomials are dense in D_α, it follows readily that the set

$$\left\{ \frac{1}{1 - \lambda z} : \lambda \in \mathbb{D} \right\}$$

has dense linear span in D_α. Furthermore, if $A \subset \mathbb{D}$ has an accumulation point in $\mathbb{D}$, and $g \in D_{-\alpha}$ annihilates

(8.4.5) $$\left\{ \frac{1}{1 - \lambda z} : \lambda \in A \right\},$$

then

$$\lambda \to \left\langle \frac{1}{1 - \lambda z}, g \right\rangle$$

is an analytic function on $\mathbb{D}$ whose zeros have an accumulation point in $\mathbb{D}$, and hence must be the zero function. By the Hahn-Banach theorem, the set in eq.(8.4.5) has dense linear span. This, together with Proposition 8.4.3, yields the following corollary.

COROLLARY 8.4.6.   (1) $\sigma_{ap}(T) \cap \mathbb{D} = \sigma_p(T) \cap \mathbb{D}$ *is either discrete or all of* $\mathbb{D}$.
(2) $\sigma_{ap}(T) = \mathbb{D}^-$ *if and only if* $\mathcal{M} = D_\alpha$.

Also notice that since $\partial \sigma(L|\mathcal{M}) \subset \sigma_{ap}(L|\mathcal{M})$ [**37**, p. 215], an elementary argument yields the following dichotomy.

COROLLARY 8.4.7. *Either*

$$\sigma(T) \cap \mathbb{D} = \sigma_{ap}(T) \cap \mathbb{D} = \sigma_p(T) \cap \mathbb{D}$$

and this set is discrete, or, $\sigma(T) = \mathbb{D}^-$.

It is important in the above corollary that we look at the portions of $\sigma(T), \sigma_{ap}(T),$ and $\sigma_p(T)$ which lie in the open unit disk $\mathbb{D}$. For $\alpha > 0$, it can be the case that $\sigma(T) = \mathbb{D}^-$ but $\sigma_{ap}(T) = \mathbb{T}$, and $\sigma_p(T) \cap \mathbb{D} = \emptyset$ [**11**, Prop. 6.1].

Suppose that $(I - \lambda T)^{-1}$ exists. Then for each $f \in \mathcal{M}$, there is a corresponding $g \in \mathcal{M}$ satisfying the equation

$$f = (I - \lambda T)g$$

or in other words,

$$g = \frac{zf - \lambda g(0)}{z - \lambda}.$$

Putting this all together, we have

(8.4.8) $$(I - \lambda T)^{-1}f = \frac{zf - \lambda c_\lambda(f)}{z - \lambda},$$

where $c_\lambda(f)$ is some complex constant depending on λ and f. For $\lambda \in \mathbb{D}$, the right-hand side of the above equation must be an analytic function (of z) on the disk and so $c_\lambda(f) = f(\lambda)$. We will see momentarily that for $\lambda \in \mathbb{D}_e$, the constant $c_\lambda(f)$ will be the value of the 'continuation' (in fact a (FM) continuation) of f at the point λ. For now, observe that

$$c_\lambda(f) = ((I - \lambda T)^{-1}f)(0)$$

is an analytic function of λ on $\{\lambda \in \mathbb{C}_\infty : 1/\lambda \notin \sigma(T)\}$ and moreover, the linear functional

$$f \to c_\lambda(f)$$

is continuous on $\mathcal{M}$.

The above discussion can be employed to characterize the part of the spectrum of T which intersects the unit circle.

PROPOSITION 8.4.9.
$$\sigma_{ap}(T) \cap \mathbb{T} = \sigma(T) \cap \mathbb{T}.$$

Moreover, $\zeta \in \mathbb{T}$ does not belong to $\sigma(T) \cap \mathbb{T}$ if and only if there is an open neighborhood U of $\bar\zeta$ such that every $f \in \mathcal{M}$ extends to be analytic in U.

PROOF. Let $\zeta \in \mathbb{T}$ with $\zeta \notin \sigma_{ap}(T)$. Since $\partial\sigma(T) \subset \sigma_{ap}(T)$ and $\sigma(T) \subset \mathbb{D}^-$, then $\zeta \notin \sigma(T)$. Thus for each $f \in \mathcal{M}$,

$$w \to ((I - wT)^{-1}f)(0)$$

is an analytic function in some neighborhood U of $\bar\zeta$. As noted earlier,

$$((I - wT)^{-1}f)(0) = f(w), \quad |w| < 1$$

and so f has an analytic continuation to U.

Now suppose $\zeta \in \mathbb{T}$ is such that there is a neighborhood U such that every $f \in \mathcal{M}$ extends to be analytic in U. The function

$$g = \frac{zf - \zeta f(\zeta)}{z - \zeta}$$

is the norm limit of

$$g_n = \frac{zf - w_n f(w_n)}{z - w_n} \in \mathcal{M},$$

for some $w_n \in \mathbb{D}$ and $w_n \to \zeta$ and so $g \in \mathcal{M}$. At least formally,

$$(8.4.10) \qquad (I - \zeta T)^{-1}f = \frac{zf - \zeta f(\zeta)}{z - \zeta}.$$

Suppose that $\{f_n\}$ is a sequence in $\mathcal{M}$ with

$$(8.4.11) \qquad f_n \to 0 \quad \text{and} \quad \frac{zf_n - \zeta f_n(\zeta)}{z - \zeta} \to h \quad \text{in norm.}$$

Note that eq.(8.4.11) also holds pointwise in $\mathbb{D}$ and so $f_n(\zeta) \to c$. Thus

$$h(z) = \frac{-c\zeta}{z - \zeta}$$

which can only hold when $c = 0$ or else h would not be analytic near ζ and would contradict our choice of ζ. Thus by the closed graph theorem, the right-hand-side of eq.(8.4.10) is a continuous operator on $\mathcal{M}$ and so $\bar{\zeta} \notin \sigma(T)$. □

So far, we have that

$$(I - \lambda T)^{-1}f = \frac{zf - \lambda c_\lambda(f)}{z - \lambda}$$

for some constant $c_\lambda(f)$. Moreover, $c_\lambda(f) = f(\lambda)$ when $\lambda \in \mathbb{D}$. As we shall see momentarily, $c_\lambda(f)$ will be the value of the (FM) continuation of $f \in \mathcal{M}$ at the point $\lambda \in \mathbb{D}_e$, at least whenever $1/\lambda \notin \sigma(T)$.

For $f \in \mathcal{M}$ and a non-trivial $g \in \mathcal{M}^\perp$, let $C_{f,g}$ be the meromorphic function defined on $\mathbb{D}_e$ by

$$C_{f,g}(\lambda) := \left\langle \frac{zf}{z - \lambda}, g \right\rangle / \left\langle \frac{\lambda}{z - \lambda}, g \right\rangle.$$

We mention a few things about this function $C_{f,g}$. The first is that

$$\left\langle \frac{\lambda}{z - \lambda}, g \right\rangle = -\bar{g}(1/\bar{\lambda}), \quad |\lambda| > 1,$$

and so the possible poles of $C_{f,g}$ are located at the reflected zeros of g. The second observation uses the fact that $\langle f, g \rangle = 0$ to write $C_{f,g}$ alternatively as

$$C_{f,g}(\lambda) = \left\langle \frac{f}{z - \lambda}, g \right\rangle / \left\langle \frac{1}{z - \lambda}, g \right\rangle.$$

This alternative formula for $C_{f,g}$, albeit a triviality, often comes in handy when proving things about $C_{f,g}$ such as the following fact.

PROPOSITION 8.4.12. *For each $f \in \mathcal{M}$ and non-trivial $g \in \mathcal{M}^\perp$, the meromorphic function $C_{f,g}$ is a (FM) continuation of f.*

PROOF. Let

$$F(\lambda) := \left\langle \frac{f}{z - \lambda}, g \right\rangle, \quad G(\lambda) := \left\langle \frac{1}{z - \lambda}, g \right\rangle, \quad \lambda \in \mathbb{D}_e$$

and note that if $f = \sum_j a_j z^j$ and $g = \sum_j b_j z^j$, then

$$
\begin{aligned}
F(\lambda) &= -\frac{1}{\lambda} \left\langle \frac{f}{1 - z/\lambda}, g \right\rangle \\
&= -\frac{1}{\lambda} \sum_{j=0}^{\infty} \langle z^j f, g \rangle \frac{1}{\lambda^j} \\
&= -\frac{1}{\lambda} \sum_{j=0}^{\infty} \left\{ \sum_{k=0}^{\infty} a_k \overline{b_{j+k}} \right\} \frac{1}{\lambda^j}.
\end{aligned}
$$

A similar calculation shows that

$$
G(\lambda) = -\frac{1}{\lambda} \sum_{j=0}^{\infty} \overline{b_j} \frac{1}{\lambda^j} = -\frac{1}{\lambda} \overline{g}(1/\overline{\lambda}).
$$

From here, one can see, using

$$
0 = \langle B^n f, g \rangle = \sum_{j=0}^{\infty} a_{n+j} \overline{b_j}, \quad n = 0, 1, 2, \ldots
$$

that $f \# G = F$ and so $F/G = C_{f,g}$ is a (FM) continuation of f for each non-trivial $g \in \mathcal{M}^{\perp}$.   □

As we shall see momentarily in § 8.5 of this chapter, one can, for certain $\alpha > 0$, choose a particularly pathological B-invariant subspace $\mathcal{M} \subset D_\alpha$ for which there is an $f \in \mathcal{M}$ and non-zero g_1, g_2 from $\mathcal{M}^{\perp}$ such that $C_{f,g_1} \neq C_{f,g_2}$. Thus C_{f,g_1} and C_{f,g_2} will be unequal (FM) continuations of f. When $\alpha = 0$, recall from eq.(6.3.6) that $C_{f,g}$ is a pseudocontinuation of f for any non-trivial annihilating g and so, by the Lusin-Privalov uniqueness theorem (Theorem 2.2.2), $C_{f,g}$ is independent of g. For a similar reason, the same is true for $\alpha < 0$ (the weighted Bergman spaces) [**11**, Thm. 4.6].

The link between the transform $C_{f,g}$ and $c_\lambda(f)$, the constant in the formula

$$
(I - \lambda T)^{-1} f = \frac{zf - \lambda c_\lambda(f)}{z - \lambda},
$$

comes with the following theorem.

PROPOSITION 8.4.13. *If $|\lambda| > 1$ with $1/\lambda \notin \sigma_{ap}(T)$ then $(I - \lambda T)^{-1}$ exists if and only if for every $f \in \mathcal{M}$, the quantity*

$$
C_{f,g}(\lambda) = \left\langle \frac{zf}{z - \lambda}, g \right\rangle \Big/ \left\langle \frac{\lambda}{z - \lambda}, g \right\rangle
$$

is independent of the choice of

(8.4.14) $g \in \mathcal{M}^{\perp}$ *with* $\left\langle \dfrac{\lambda}{z - \lambda}, g \right\rangle = -\overline{g}(1/\overline{\lambda}) \neq 0.$

In fact, if $|\lambda| > 1$ with $1/\lambda \notin \sigma(T)$, then $C_{f,g}(\lambda) = c_\lambda(f)$ and

$$
(I - \lambda T)^{-1} f = \frac{zf - \lambda c_\lambda(f)}{z - \lambda}
$$

for all $f \in \mathcal{M}$ and g satisfying eq.(8.4.14).

PROOF. Suppose $|\lambda| > 1$ with $1/\lambda \notin \sigma(T)$. Then, as argued to prove eq.(8.4.8), for each $f \in \mathcal{M}$ there exists a constant $c_\lambda(f) \in \mathbb{C}$ such that

$$(I - \lambda T)^{-1} f = \frac{zf - \lambda c_\lambda(f)}{z - \lambda}.$$

Now let g satisfy eq.(8.4.14). Notice that by Proposition 8.4.3, $(z - \lambda)^{-1} \notin \mathcal{M}$ and so such g's indeed exist. We have

$$0 = \langle (I - \lambda T)^{-1} f, g \rangle = \langle \frac{zf}{z - \lambda}, g \rangle - c_\lambda(f) \langle \frac{\lambda}{z - \lambda}, g \rangle$$

and so

$$c_\lambda(f) = \langle \frac{zf}{z - \lambda}, g \rangle / \langle \frac{\lambda}{z - \lambda}, g \rangle = C_{f,g}(\lambda)$$

is independent of g.

For the other direction, let $|\lambda| > 1$ with $1/\lambda \notin \sigma_{ap}(T)$. Thus, as above, we may choose g satisfying eq.(8.4.14). For $f \in D_\alpha$, define

$$R_\lambda f = \frac{zf - \lambda C_{f,g}(\lambda)}{z - \lambda}$$

and note that $R_\lambda D_\alpha \subset D_\alpha$. Furthermore, an application of the closed graph theorem shows that R_λ is continuous. Since $1/\lambda \notin \sigma_{ap}(T)$, at least formally, R_λ is an inverse of $I - \lambda T$. What needs to be shown is $R_\lambda \mathcal{M} \subset \mathcal{M}$. To this end, let $h \in \mathcal{M}^\perp$ and notice that since $\langle (z - \lambda)^{-1}, g \rangle \neq 0$, there is a sequence of complex numbers $a_n \to 0$ such that

$$h - a_n g \in \mathcal{M}^\perp \quad \text{and} \quad \langle \frac{1}{z - \lambda}, h - a_n g \rangle \neq 0.$$

Then

$$
\begin{aligned}
\langle R_\lambda f, h - a_n g \rangle &= \langle \frac{zf}{z - \lambda}, h - a_n g \rangle - C_{f,g}(\lambda) \langle \frac{\lambda}{z - \lambda}, h - a_n g \rangle \\
&= \langle \frac{zf}{z - \lambda}, h - a_n g \rangle - C_{f,h-a_n g}(\lambda) \langle \frac{\lambda}{z - \lambda}, h - a_n g \rangle \\
&= 0
\end{aligned}
$$

since, by assumption, $C_{f,g}(\lambda)$ is independent of g. We now let $a_n \to 0$ and find that $\langle R_\lambda f, h \rangle = 0$ for arbitrary $h \in \mathcal{M}^\perp$, implying $R_\lambda \mathcal{M} \subset \mathcal{M}$. $\square$

Recall that for a B-invariant subspace $\mathcal{M} \subset D_\alpha$, the annihilator $\mathcal{M}^\perp$ is an S-invariant subspace of $D_{-\alpha}$. As it turns out, there is a beautiful connection between $\sigma(T) \cap \mathbb{D}$ and a certain property of $\mathcal{M}^\perp$ called the 'division property'. We outline some of the basic facts about the division property here and refer the reader to a paper of Richter [111] for further details.

DEFINITION 8.4.15. For a non-trivial S-invariant subspace $\mathcal{K}$ of D_α, define the 'index' of $\mathcal{K}$ to be the number

$$\text{ind}(\mathcal{K}) := \dim(\mathcal{K}/S\mathcal{K}).$$

EXAMPLE 8.4.16. (1) If

$$[f] := \bigvee \{S^n f : n = 0, 1, 2, \ldots\}$$

is the S-invariant subspace generated by a single non-trivial function $f \in D_\alpha$, then it is not difficult to show that $\text{ind}([f]) = 1$.

(2) If $\{a_n\}$ is a sequence of points in the disk which is the zero set of some non-trivial function in D_α, then the (non-zero) subspace

$$\{f \in D_\alpha : f(\{a_n\}) = 0\}$$

is a closed S-invariant subspace with index equal to one.

(3) For $\alpha \geq 0$, *every* (non-zero) S-invariant subspace of D_α has index equal to one. For $\alpha = 0$, this is a consequence of Beurling's theorem (Theorem 6.3.1). The $\alpha = 1$ case was done by Richter and Shields [**112**] while the $0 < \alpha < 1$ case was done by Aleman [**9**]. When $\alpha > 1$, D_α is an algebra and the result follows from a theorem of Bourdon [**29**].

(4) Unlike the case when $\alpha \geq 0$, where every non-trivial S-invariant subspace of D_α has index equal to one, it is also known [**17, 70, 72**] that given $\alpha < 0$ and any integer $n \in \mathbb{N} \cup \{\infty\}$, there is an S-invariant $\mathcal{K}$ of D_α such that $\mathrm{ind}(\mathcal{K}) = n$.

S-invariant subspaces with index equal to one are important and can be classified in a variety of equivalent ways. One way uses the 'division property'.

DEFINITION 8.4.17. We say a non-trivial S-invariant subspace $\mathcal{K} \subset D_\alpha$ has the 'division property' if for each $\lambda \in \mathbb{D} \backslash Z(\mathcal{K})$, where

$$(8.4.18) \qquad Z(\mathcal{K}) := \bigcap_{f \in \mathcal{K}} f^{-1}(\{0\})$$

is the set of common zeros of $\mathcal{K}$, we have

$$(8.4.19) \qquad \frac{f}{z - \lambda} \in \mathcal{K} \quad \text{for all } f \in \mathcal{K} \text{ with } f(\lambda) = 0.$$

By Fredholm theory [**111**, Lemma 2.1], if eq.(8.4.19) holds for a single $\lambda \in \mathbb{D} \backslash Z(\mathcal{K})$, it holds for all $\lambda \in \mathbb{D} \backslash Z(\mathcal{K})$. The following lemma [**111**, Lemma 3.1] relates the index and the division property.

THEOREM 8.4.20. *For a non-trivial S-invariant subspace $\mathcal{K} \subset D_\alpha$, the following are equivalent.*

(1) *$\mathcal{K}$ has index equal to one.*

(2) *$\mathcal{K}$ has the division property.*

For a B-invariant subspace $\mathcal{M}$ of D_α, recall from Proposition 8.4.3 that

$$\sigma_{ap}(T) \cap \mathbb{D} = \sigma_p(T) \cap \mathbb{D} = \Big\{ a \in \mathbb{D} : \frac{1}{1 - az} \in \mathcal{M} \Big\}.$$

From the identity

$$\Big\langle \frac{1}{1 - az}, g \Big\rangle = \overline{g}(\overline{a}),$$

we also have the following corollary.

COROLLARY 8.4.21. $\sigma_{ap}(T) \cap \mathbb{D} = \sigma_p(T) \cap \mathbb{D} = \{\overline{\lambda} : \lambda \in Z(\mathcal{M}^\perp)\}$.

The following result of Richter [**111**, Thm. 4.5] relates $\sigma(T)$ to the division property.

THEOREM 8.4.22 (Richter). *Let $\mathcal{M}$ be a non-trivial B-invariant subspace of D_α.*

(1) *If $\mathcal{M}^\perp$ has the division property (equivalently, $ind(\mathcal{M}^\perp) = 1$), then*

$$\sigma_{ap}(T) \cap \mathbb{D} = \sigma(T) \cap \mathbb{D} = \{\bar{\lambda} : \lambda \in Z(\mathcal{M}^\perp)\}.$$

(2) *If $\mathcal{M}^\perp$ does not have the division property (equivalently, $ind(\mathcal{M}^\perp) > 1$), then $\sigma(T) = \mathbb{D}^-$.*

REMARK 8.4.23. The results of this section on (FM) continuation hold for a large class of Banach space of analytic functions on $\mathbb{D}$ and we leave it to the reader to check that all the proofs hold *mutatis mutandis* for a general Banach space $\mathcal{X}$ of analytic functions on $\mathbb{D}$ which satisfy the conditions

(1) The inclusion operator $i : \mathcal{X} \to \mathfrak{H}(\mathbb{D})$ is injective and continuous.
(2) $1 \in \mathcal{X}$.
(3) The shift operator $Sf = zf$ is continuous on $\mathcal{X}$ with $\sigma(S) = \mathbb{D}^-$.
(4) For all $\lambda \in \mathbb{D}$,

$$\frac{f - f(\lambda)}{z - \lambda} \in \mathcal{X}$$

whenever $f \in \mathcal{X}$.

(5) If $f \in \mathcal{X}$ is analytic in a neighborhood of a boundary point $\zeta \in \mathbb{T}$, then

$$\frac{f - f(w)}{z - w} \to \frac{f - f(\zeta)}{z - \zeta}$$

in the norm of $\mathcal{X}$ as $w \to \zeta$ ($w \in \mathbb{D}$).

(6) The dual, $\mathcal{X}^*$, of $\mathcal{X}$ can be equated with a Banach space of analytic functions on $\mathbb{D}$ by means of the Cauchy pairing

$$\langle f, g \rangle = \sum_{n=0}^\infty a_n \overline{b_n},$$

where $\{a_n\}$ are the Taylor coefficients of $f \in \mathcal{X}$ and $\{b_n\}$ are those of $g \in \mathcal{X}^*$.

(7) With f and g as above, the technical hypotheses of Theorem 8.2.1 hold.

Certainly any of the D_α spaces satisfy the above conditions as do the ℓ_A^p spaces (see Remark 8.3.13). We also refer the reader to the paper [11] where they examine the spectral properties of $T = B|\mathcal{M}$ for various other spaces of analytic functions.

8.5. Uniqueness

If f belongs to a non-trivial B-invariant subspace $\mathcal{M}$ of D_α and g is a non-trivial function in $\mathcal{M}^\perp$, then

$$C_{f,g}(\lambda) := \langle \frac{f}{z - \lambda}, g \rangle / \langle \frac{1}{z - \lambda}, g \rangle$$

is meromorphic on $\mathbb{D}_e$ and is also a (FM) continuation of f (Proposition 8.4.12). In this section, we wish to show the possible dependence of $C_{f,g}$ on the annihilating g, thus showing that (FM) continuations are not always unique. Although this was essentially laid out in the previous section, we formalize things here.

When $\alpha = 0$ (the Hardy space case), $C_{f,g}$ is the pseudocontinuation of f and so, since pseudocontinuations are unique, $C_{f,g}$ is independent of the annihilating g. When $\alpha < 0$ (the weighted Bergman spaces), the same is true [**11**, Thm. 4.6]. When $\alpha > 0$, the situation is different, and, as discussed in Example 8.4.16, it is possible for the S-invariant subspace $\mathcal{M}^{\perp}$ to not have the division property. From Theorem 8.4.22 and Proposition 8.4.13 we obtain the following result.

THEOREM 8.5.1. *Let $\mathcal{M}$ be a B-invariant subspace of $D_{\alpha}(\alpha > 0)$ satisfying the condition $\mathrm{ind}(\mathcal{M}^{\perp}) > 1$. Then there is an $f \in \mathcal{M}$ and non-trivial $g_1, g_2 \in \mathcal{M}^{\perp}$ such that $C_{f,g_1} \neq C_{f,g_2}$.*

We have seen, through the Borel series example in Remark 8.3.5, an example of a function f which has a (FM) continuation which is not a pseudocontinuation of f. We have also seen above that a given function might have two different (FM) continuations. Next, we give an example of a function f which has two different (FM) continuations. One will be a pseudocontinuation of f while the other will not. For our example of this, we will focus on the duality between D_1 (the classical Dirichlet space) and D_{-1} (the Bergman space). Recall from Proposition 8.4.12 that for any non-trivial $g \in D_{-1}$ and f in $[g]^{\perp}$, the annihilator of $[g] = \bigvee\{z^n g : n = 0, 1, 2, \ldots\}$, the meromorphic function

$$\lambda \to C_{f,g}(\lambda) = \langle \frac{zf}{z - \lambda}, g \rangle \,/\, \langle \frac{\lambda}{z - \lambda}, g \rangle = \langle \frac{zf}{z - \lambda}, g \rangle \,/\, -\overline{g}(1/\overline{\lambda})$$

is a (FM) continuation of f. This next technical lemma can be found in [**11**, Lemma 7.4].

LEMMA 8.5.2. *Let $g \in D_{-1}$ and $f \in [g]^{\perp} \subset D_1$. Then for almost every $e^{i\theta}$,*

$$\overline{g}(1/\overline{\lambda}) \left\{ C_{f,g}(\lambda) - f(e^{i\theta}) \right\} \to 0$$

as $\lambda \to e^{i\theta}$, $|\lambda| > 1$, non-tangentially.

This lemma has two interesting corollaries. The first follows from the previous lemma and the Lusin-Privalov uniqueness theorem (Theorem 2.2.2).

COROLLARY 8.5.3. *Let $g \in D_{-1}$ and $f \in [g]^{\perp} \subset D_1$. If $C_{f,g}$ has finite non-tangential limits almost everywhere on a non-degenerate arc $I \subset \mathbb{T}$, then $C_{f,g}$ is a pseudocontinuation of f across I.*

COROLLARY 8.5.4. *If $g \in D_{-1}$ has finite non-tangential limits almost everywhere on a non-degenerate arc $I \subset \mathbb{T}$, then for any $f \in [g]^{\perp}$, the function $C_{f,g}$ is a pseudocontinuation of f across I.*

EXAMPLE 8.5.5. We are now ready to give an example of a function f which has two different (FM) continuations. One will be a pseudocontinuation of f while the other will not. By a recent result of Aleman, Richter, and Sundberg [**12**], there is a non-trivial $g \in D_{-1}$ which has finite non-tangential limits almost everywhere on $\mathbb{T}$ and belongs to some S-invariant subspace $\mathcal{K}$ of D_{-1} with index equal to 2. From Corollary 8.5.4, $C_{f,g}$ is a pseudocontinuation of f for all $f \in \mathcal{K}^{\perp} \subset [g]^{\perp}$. Since $\mathcal{K}$ does not have the division property, it will follow from Theorem 8.4.22 and

Proposition 8.4.13 the existence of an $f \in \mathcal{K}^\perp$ and a $g_1 \in \mathcal{K}$ such that $C_{f,g}$ does not equal C_{f,g_1}. From Proposition 8.4.12, both $C_{f,g}$ and C_{f,g_1} are (FM) continuations of f.

8.6. A continuation arising from overconvergence

As we have seen previously, the non-cyclic vectors for the backward shift on D_α have 'continuations' (of one sort or another) to meromorphic functions on the exterior disk. For the Hardy space $D_0 = H^2$, the non-cyclic vectors have pseudocontinuations of bounded type (Theorem 6.3.4). For the Bergman-type spaces D_α ($\alpha < 0$), the same is true (Theorem 6.6.1), although, unlike the Hardy space, the existence of a pseudocontinuation of bounded type does not always imply non-cyclicity (see Chapter 6, § 6.6). When $\alpha > 0$, the non-cyclic vectors in D_α need not have pseudocontinuations across any portion of the circle (see eq.(6.6.6)), however they do have a (FM) continuation (Proposition 8.3.4).

In this section, we examine this 'continuation' phenomenon further for a general abstract Banach space of analytic functions on $\mathbb{D}$. In this general setting, one can imagine that not much can be said. However, if we focus on the properties of the vectors which are contained in backward shift invariant subspaces satisfying a 'spectral synthesis property' (Definition 6.5.3), much can be said, even in the general abstract setting.

Our set up is as follows: Consider a Banach space $\mathcal{X}$ of analytic functions on $\mathbb{D}$ satisfying the conditions

(1) $1 \in \mathcal{X}$ and $\dfrac{1}{z-b} \in \mathcal{X}$ for all $|b| > 1$. Moreover,

$$\bigvee \left\{ \frac{1}{z-b} : |b| > 1 \right\} = \mathcal{X}.$$

(2) For each $a \in \mathbb{D}$, the evaluation functional $I_a f = f(a)$ is continuous on $\mathcal{X}$.

(3) $\lim\limits_{b \to \infty} \left\| \dfrac{1}{z-b} \right\| = 0$.

(4) For each $|b| > 1$, $M_b : \mathcal{X} \to \mathcal{X}$,

$$M_b f = \frac{f}{z-b}$$

is defined and continuous.

(5) $\|M_b\|$ is bounded on compact subsets of $\mathbb{D}_e$.

(6) For some $1 < p < \infty$, the following two conditions hold.

$$\int_0^1 \int_0^{2\pi} (\log^+ \log^+ \|I_a\|)^p r \, dr \, dt < \infty, \quad a = re^{it}$$

$$\int_0^Q (\log^+ \log^+ N(x))^p dx < \infty$$

for every $1 < Q < \infty$. Here $N(x)$ satisfies $\|M_b\| < N(|b|)$ for $|b| > 1$.

Certainly all of the D_α spaces satisfy these conditions as well as a host of others such as ℓ_A^p ($1 < p < \infty$).

Notice in the above set up, there was no mention of a backward shift operator on $\mathcal{X}$. We will include it later. Let E be a set of points in $\mathbb{D}_e$ such that the set $\mathcal{R}_E$, of all finite linear combinations of the functions

$$\frac{1}{z-b}, \quad b \in E,$$

is not dense in $\mathcal{X}$. For this to hold, notice that E can have no accumulation points in $\mathbb{D}_e$ (see eq.(8.4.5)). If $f \in \mathcal{R}_E^-$, then f is the norm limit of a sequence $\{f_n\} \subset \mathcal{R}_E$ and so by condition (2) above, this sequence converges uniformly on compact subsets of $\mathbb{D}$ to f. The following 'overconvergence' result, which we state without proof, says quite a bit more.

THEOREM 8.6.1 (Shapiro [**129**]). *With the above notation, if a sequence of rational functions $\{f_n\} \subset R_E$ converges in norm, then the sequence $\{f_n\}$ converges uniformly on compact subsets of $\mathbb{C}_\infty \backslash E^-$.*

Thus each function $f \in \mathcal{R}_E^-$ can be associated with a function $S_f \in \mathfrak{H}(\mathbb{D}_e \backslash E^-)$ and so, in some sense, S_f is a 'continuation' of f.

QUESTION 8.6.2. With this 'overconvergence' continuation, there are several unanswered questions. (i) From the proof of the above theorem, one can show that $f \equiv 0$ implies $S_f \equiv 0$ and so f uniquely determines S_f. If $S_f \equiv 0$, does this imply $f \equiv 0$? (ii) Is there an 'Abelian theorem' for this type of continuation? That is to say, if f has an analytic continuation across some boundary point $e^{i\theta}$, must this continuation coincide with S_f?

Returning to the Dirichlet type spaces D_α, or more generally to a Banach space $\mathcal{X}$ as above which *also* satisfies the conditions outlined in Remark 8.4.23, we see that the vectors $\{1/(z-b) : b \in E\}$, are eigenvectors for the backward shift B and so the $\mathcal{R}_E^-$ is a non-trivial B-invariant subspace of $\mathcal{X}$ which satisfies the 'spectral synthesis property' (Definition 6.5.3). By Proposition 8.3.4 and Proposition 8.4.12, every $f \in \mathcal{R}_E^-$ has a (FM) continuation T_f given by

$$T_f(\lambda) = \langle \frac{f}{z-\lambda}, g \rangle \, / \, \langle \frac{1}{z-\lambda}, g \rangle, \quad |\lambda| > 1,$$

where $g \in \mathcal{R}_E^\perp \backslash (0)$. Moreover, since the S-invariant subspace

$$\mathcal{R}_E^\perp = \{ g \in \mathcal{X}^* : g(1/\overline{b}) = 0 \text{ for all } b \in E \}$$

has the division property (Example 8.4.16), we see from Theorem 8.4.22 and Proposition 8.4.13 that the above definition of T_f is independent of the annihilating g.

So, we seem to have two 'continuations' associated to a function $f \in \mathcal{R}_E^-$: the continuation S_f arising from the overconvergence of a sequence $\{f_n\} \subset \mathcal{R}_E$ approximating f in norm, and the (FM) continuation T_f. Are these the same? Fortunately, the answer is yes and can be seen as follows: Let $\{f_n\} \subset \mathcal{R}_E$ approximate f in norm. By Theorem 8.2.1, (FM) continuation is compatible with analytic continuation and so (since each f_n is a rational function)

$$f_n(\lambda) = S_{f_n}(\lambda) = T_{f_n}(\lambda) = \langle \frac{f_n}{z-\lambda}, g \rangle \, / \, \langle \frac{1}{z-\lambda}, g \rangle, \quad |\lambda| > 1.$$

By the above overconvergence theorem, $S_{f_n} \to S_f$ uniformly on compact subsets of $\mathbb{C}_\infty \backslash E^-$. But since multiplication by $1/(z-\lambda)$ is continuous on $\mathcal{X}$ and $f_n \to f$ in norm, then $T_{f_n} \to T_f$ uniformly on compact subsets of $\mathbb{D}_e$. Thus $T_f = S_f$.

8.7. Generalizing the Walsh-Tumarkin results

Using the notation from Theorem 6.5.5, *i.e.*,

$$(8.7.1) \qquad S_n := \{ z_{n,1}, \dots, z_{n,N(n)} \} \subset \mathbb{D}, \quad n, N(n) \in \mathbb{N}$$

and

$$(8.7.2) \qquad R_n := \bigvee_{j=1}^{N(n)} \left\{ k_{z_{n,j}}^{(s)}, j = 1, \dots, N(n), s = 0, 1, \dots, \mathrm{mult}(z_{n,j}) - 1 \right\},$$

where

$$k_\lambda^{(n)}(z) = \frac{n! z^n}{(1 - \bar{\lambda} z)^{n+1}},$$

one can attempt to generalize the Walsh-Tumarkin results to the D_α spaces. In particular, we focus on the following questions.

QUESTION 8.7.3.   (1) Under what conditions is

$$\varliminf R_n := \left\{ f \in D_\alpha : \lim_{n \to \infty} \mathrm{dist}(f, R_n) = 0 \right\}$$

equal to D_α (the Walsh question)?
  (2) When

$$\varliminf R_n \neq D_\alpha,$$

do the functions in $\varliminf R_n$ have any 'continuation' properties to the exterior disk (the Tumarkin question)?
  (3) These 'liminf' spaces are B-invariant subspaces of D_α. Is *every* B-invariant subspace of D_α of this form (*i.e.*, admit 'approximate spectral synthesis')?

To examine our first question, Walsh showed (Theorem 6.1.2) for the Hardy space $H^2 = D_0$, that

$$\varliminf R_n \neq H^2 \Leftrightarrow \varliminf_{n \to \infty} \mathrm{cap}(R_n) < \infty,$$

where

$$(8.7.4) \qquad \mathrm{cap}(R_n) := \sum_{j=1}^{N(n)} (1 - |z_{n,j}|).$$

Gribov and Nikol'skiĭ [64] (see also [103, p. 39]) as well as Hilden and Wallen [73] replace the sum in eq.(8.7.4) with a suitable 'capacity' cap_α for the D_α spaces and prove the following Walsh-type theorem.

THEOREM 8.7.5 (Gribov - Nikol'skiĭ).

$$\varliminf R_n \neq D_\alpha \Leftrightarrow \varliminf_{n \to \infty} \mathrm{cap}_\alpha(R_n) < \infty.$$

The above result is true in a more general setting.

For our second question, we recall from Tumarkin's theorem (Theorem 6.1.3) that when

$$\varliminf_{n \to \infty} \mathrm{cap}(R_n) < \infty,$$

then every $f \in \varliminf R_n$ has a pseudocontinuation of bounded type in $\mathbb{D}_e$, or from Definition 6.3.3, $f \in PCBT$. Moreover, for every $f \in PCBT$, there is a tableau

$\{S_n : n = 1, 2, \ldots\}$ with $\text{cap}(R_n)$ bounded and a sequence $\{f_n : f_n \in R_n\}$ such that $f_n \to f$ in H^2 norm.

For the D_α spaces, what type of 'continuations' do functions from $\varprojlim R_n$ have? Clearly this subspace is a non-trivial B-invariant subspace of D_α and hence every element is a non-cyclic vector. From Proposition 8.3.4, every element of $\varprojlim R_n$ has a (FM) continuation of a certain type. Furthermore, as we have seen from Theorem 8.6.1 and the remarks following, for certain B-invariant subspaces - those which satisfy a spectral synthesis property (Definition 6.5.3) - there is an 'overconvergence' phenomenon which gives rise to a 'continuation', which, fortunately, is the same as the (FM) continuation. Our main theorem of this section tries to generalize this overconvergence result (Theorem 8.6.1) beyond the spectral synthesis property to a Walsh-Tumarkin tableau situation.

THEOREM 8.7.6. *Let S_n and R_n be as in eq.(8.7.1) and eq.(8.7.2) and assume that $R := \varprojlim R_n \neq D_\alpha$. Let $S_n^* = \{1/\bar{z} : z \in S_n\}$, $F_m := (\bigcup_{n \geq m} S_n^*)^-$, and $F := \bigcap_m F_m$.*

(1) *If the sequence $\{f_n : f_n \in R_n\}$ is norm convergent to $f \in R$, then given any $w \in \mathbb{D}_e \backslash F$, there is an open set U_w containing w and a subsequence $\{f_{n_j}\}$ which forms a normal family on U_w.*

(2) *Every $f \in R$ has a (FM) continuation given by the formula*

$$C_{f,g}(\lambda) := \langle \frac{f}{z - \lambda}, g \rangle \, / \, \langle \frac{1}{z - \lambda}, g \rangle$$

for some $g \in R^\perp$.

(3) *If (by passing to a subsequence if necessary) we assume that the subsequence $\{f_{n_j}\}$ in statement (1) converges uniformly on compact subsets of U_w, then there is a $g \in R^\perp$ such that $f_{n_j} \to C_{f,g}$ uniformly on compact subsets of U_w.*

REMARK 8.7.7. Again we emphasize that, at least locally, the 'continuation' of $f \in \varprojlim R_n$ coming from this weak sort of overconvergence is equal to *some* (FM) continuation of f.

PROOF. Notice that statement (2) of the theorem follows from Proposition 8.4.12. We now prove statement (1). Let $w \in \mathbb{D}_e \backslash S_n^*$ and, let $\ell \in D_\alpha^*$ with $\|\ell\| = 1$, $\ell(R_n) = 0$, and

$$\ell(\frac{1}{z - w}) = d_n(w) := \text{dist}(\frac{1}{z - w}, R_n).$$

The existence of ℓ is assured by the Hahn-Banach theorem. For $a \in S_n^* \backslash \{\infty\}$ (We assume, for simplicity, that all points of S_n^* have multiplicity one. The proof can be suitably modified if this is not the case.), we have

$$\frac{1}{z - w} - \frac{1}{z - a} = \frac{w - a}{(z - w)(z - a)}.$$

Applying ℓ to both sides yields

$$d_n(w) = (w - a)\ell \left(\frac{1}{(z - w)(z - a)} \right),$$

or equivalently,

$$\ell \left(\frac{1}{(z-w)(z-a)} \right) = \frac{d_n(w)}{w-a}.$$

Replacing a in turn by

$$1/\overline{z_{n,1}}, \ldots, 1/\overline{z_{n,N(n)}}$$

and taking linear combinations of the above equations gives us

$$\ell \left(\frac{f_n}{z-w} \right) = d_n(w) f_n(w)$$

for arbitrary $f_n \in R_n$. Hence

$$(8.7.8) \qquad |f_n(w)| \leq \frac{1}{d_n(w)} \|\ell\| C_w \|f_n\| = \frac{1}{d_n(w)} C_w \|f_n\|,$$

where C_w is the operator norm of $g \to g/(z-w)$ on D_α.

Since $R \neq D_\alpha$, then $E = \{b \in \mathbb{D}_e : 1/(z-b) \in R\}$ must be a sequence with no accumulation points in $\mathbb{D}_e$. It follows from the definition of a 'liminf' space that for fixed $w \in \mathbb{D}_e \backslash (E \cup F)$, $d_n(w) \nrightarrow 0$ as $n \to \infty$. Thus there is a subsequence $\{n_j\}$ for which $d_{n_j} > c$ for some $c > 0$. Since d_{n_j} is a continuous (even Lipschitz) function, we can find a neighborhood U_w of w for which $d_{n_j} > c/2$ on U_w. From eq.(8.7.8) it follows that $\{f_{n_j}\}$ is a normal family on U_w, proving statement (1) of the theorem.

Let $w \in \mathbb{D}_e \backslash F$ and for each n, choose $g_n \in R_n^\perp$ with $\|g_n\|_{D_{-\alpha}} = 1$ and

$$\left\langle \frac{1}{z-w}, g_n \right\rangle = d_n(w)$$

(as above). Note that by passing to a subsequence if necessary, g_n converges weakly to some $g \in D_{-\alpha}$ which, since (by passing to another subsequence if necessary) $d_{n_j}(w)$ is bounded away from zero, will not be the zero function. If f is an arbitrary element of R and $f_{n_j} \in R_{n_j}$ converges in norm to f, then

$$|\langle f, g \rangle| = \lim_{n_j \to \infty} |\langle f - f_{n_j}, g_{n_j} \rangle| \leq \overline{\lim_{n_j \to \infty}} \|f - f_{n_j}\|_{D_\alpha} \|g_{n_j}\|_{D_{-\alpha}} = 0.$$

Thus $g \in R^\perp \backslash \{0\}$.

Let us assume now that f_{n_j} also converges uniformly on compact subsets of U_w. By Proposition 8.4.12, $C_{f_{n_j}, g_{n_j}}$ is a (FM) continuation of f_{n_j} and, since f_{n_j} is a rational function, Theorem 8.2.1 says $f_{n_j} = C_{f_{n_j}, g_{n_j}}$ on $\mathbb{D}_e$. Is it now a routine exercise to show that $f_{n_j}(\lambda) = C_{f_{n_j}, g_{n_j}}(\lambda) \to C_{f,g}(\lambda)$ pointwise on U_w. This proves statement (3) of the theorem. $\qquad\qquad\qquad\qquad\qquad\qquad\qquad\qquad \square$

If we replace the 'liminf' space above by

$$(8.7.9) \qquad\qquad R := \{ f \in D_\alpha : \inf_n \operatorname{dist}(f, R_n) = 0 \},$$

and assume that $R \neq D_\alpha$, then Theorem 8.7.6 can be strengthened to the following result. We leave it to the reader to make the minor modifications of the proof of the previous result to obtain this one.

THEOREM 8.7.10. *If $R \neq D_\alpha$ be as in eq.(8.7.9), $f \in R$, and $f_{n_j} \to f$, where $f_{n_j} \in R_{n_j}$. Then for each $g \in R^\perp \backslash \{0\}$, f_{n_j} converges uniformly on compact subsets of $\mathbb{D}_e \backslash F$ to $C_{f,g}$ and hence $C_{f,g}$ is independent of the choice of g.*

REMARK 8.7.11. From Theorem 8.5.1, there are B-invariant subspaces of D_α, $\alpha > 0$, for which $C_{f,g}$ depends on the choice of g in the annihilator. Thus not every B-invariant subspace can be written as in eq.(8.7.9).

Returning to the third question raised in Question 8.7.3, recall that for the Hardy space H^2 (or rather D_0), *every* B-invariant subspace $\mathcal{M}$ can be written as

$$\mathcal{M} = \underline{\lim} \; R_n.$$

A recent remarkable result of Shimorin [**133**], which uses quite deep results about the S-invariant subspaces of the Bergman space, says the same is true for the classical Dirichlet space D_1. Moreover, he also obtains an analogous condition for cyclicity (to the Tumarkin result for the Hardy space) in the Dirichlet space, namely, $f \in D_1$ is non-cyclic if and only if

$$f = \lim_{n \to \infty} f_n, \quad f_n \in R_n, \quad \sup_n \mathrm{cap}_1(R_n) < \infty.$$

8.8. Possible generalizations

Although there are cases where continuation by formal multiplication of series (FM) is applicable without postulating the existence of boundary values for the functions involved (and so pseudocontinuation is inapplicable), there are also severe limitations built into our definition of FM which make it less flexible than pseudocontinuation, namely: (a) the requirement that the relevant functions be given in disks; (b) an asymmetry between the kinds of functions to be continued: the functions being continued are required to be *holomorphic* in a disk, while their continuations are, in general, *meromorphic* in the exterior of the disk; (c) the lack of localization: a function either has a (FM) continuation across the whole circle, or it has none at all. Moreover, the continuation, when it exists, is meromorphic in all of $\mathbb{D}_e$.

It seems plausible that these restrictions might to some extent be relaxed. We leave this as a program for the future, but include some remarks about such a program. As to (a), it is not clear much more can be done. One could, for example, work with complementary half-planes instead of $\mathbb{D}$ and $\mathbb{D}_e$, and use Laplace transforms in place of power series, or perhaps invoke conformal mapping in some way. As to (b) however, there is a natural way to associate a meromorphic function $h = f/g$ in $\mathbb{D}$, where $f, g \in \mathfrak{H}(\mathbb{D})$, with one, say $H = F/G$ in $\mathbb{D}_e$, where $F, G \in \mathfrak{H}(\mathbb{D}_e)$. Indeed, if the Taylor coefficients of the functions f, g, F, and G are suitably restricted, one can require that the formal trigonometric series on $\mathbb{T}$ corresponding to the formal products fG and gF be identical. It would be interesting to see if a version of Theorem 8.2.1 could be proved for this more general scheme so that the original theorem would be the case $g \equiv 1$. As to (c), one kind of localization seems plausible: continuation across a sub-arc of $\mathbb{T}$. Indeed, under suitable assumptions, $fG - gF$ is identified with a formal trigonometric series, which may be considered as a distribution on $\mathbb{T}$. We may then call f/g and F/G 'continuations' of one another across an open sub-arc J of $\mathbb{T}$ if the support of this distribution does not meet J. Again, there is work to be done in extending Theorem 8.2.1 to this set-up.

On the other hand, it is not evident that there is a sensible generalization of FM when the initial function f/g is meromorphic only in some proper sub-domain of $\mathbb{D}$, or, perhaps, a continuation F/G is sought that is required to be meromorphic only on some proper sub-domain of $\mathbb{D}_e$, even if the two domains have a common boundary arc on $\mathbb{T}$. At bottom this difficulty is the same as in (a). The general picture is the following: we have meromorphic functions f/g and F/G in adjacent domains Ω_1 and Ω_2 which have a common boundary arc J. We want to propose these as candidates to be 'continuations' of one another when, in some suitable sense, $fG = gF$ holds on J. In the case where J is a smooth arc and the functions f, g, F, and G have nontangential boundary values almost everywhere on J, with respect to arc length measure, the last equation makes sense for the boundary values almost everywhere on J and we are in the situation of pseudocontinuation. In certain other cases, the equation $fG = Fg$ can, as we have seen, be interpreted using formal multiplication of trigonometric series on J. Perhaps there are yet other ways to interpret this equation which will lead to new notions of continuation freed from the geometric restrictions required for FM.

CHAPTER 9

Generalized analytic continuation

9.1. Axiomatizing the notion of GAC

Returning to our theme of comparing GAC with summability methods for divergent series, we recall the various ways in which one can assign a number s to represent the 'sum' $a_0 + a_1 + \cdots$ as proposed by Abel, Borel, Cauchy, Cesàro, and others (see Hardy's book [**68**, Ch. 1]). Following Hardy, we set $s = \sum a_n$ to be the 'sum' of the sequence $\{a_n\}$, according to some fixed summation method $\sum$, and hope that $\sum$ has certain 'reasonable' properties such as:

$$\text{If } \sum a_n = s, \text{ then } \sum ka_n = ks.$$

$$\text{If } \sum a_n = s \text{ and } \sum b_n = t, \text{ then } \sum(a_n + b_n) = s + t.$$

$$\text{If } \sum\nolimits_{n\geq 0} a_n = s, \text{ then } \sum\nolimits_{n\geq 1} a_n = s - a_0 \text{ and conversely.}$$

$$\text{If the series } a_0 + a_1 + a_2 + \cdots \text{ converges to } s, \text{ then } \sum a_n = s.$$

Most of the above mentioned summation methods satisfy these conditions.

Just as those early mathematicians proposed various types of summation methods which satisfy certain natural properties, in these notes, we have discussed several types of 'continuations' of certain functions $f \in \mathfrak{M}(\mathbb{D})$ to an associated function $C_f \in \mathfrak{M}(\mathbb{D}_e)$. With both Borel-Walsh (Chapter 4) and Gončar continuation (Chapter 5), we associated to each function f in a certain subset of $\mathfrak{M}(\mathbb{D})$ with a element $C_f \in \mathfrak{M}(\mathbb{D}_e)$. This was based on certain criteria of approximability of f by rational functions, superconvergence in the Borel-Walsh case, hyperconvergence in the Gončar case. With the Walsh-Tumarkin model (Chapter 6), we associated to each $f \in H^2$ which is the H^2 limit of a sequence of rational functions whose poles are restricted by a tableau subject to the condition in Theorem 6.1.3, with the pseudocontinuation $C_f \in \mathfrak{N}(\mathbb{D}_e)$ via matching non-tangential limits almost everywhere. Likewise, in the Bochner-Bohnenblust set up (Chapter 7), to each function

$$f_A = \sum_{n=0}^{\infty} A(n)z^n, \quad z \in \mathbb{D},$$

where $\{A(n) : n \in \mathbb{Z}\}$ is an almost periodic sequence, we associated the function

$$F_A = -\sum_{n=1}^{\infty} \frac{A(-n)}{z^n}, \quad z \in \mathbb{D}_e.$$

In each of these cases, we have created a well defined mapping

$$f \to C_f$$

from a set $S_1 \subset \mathfrak{M}(\mathbb{D})$ to a set $S_2 \subset \mathfrak{M}(\mathbb{D}_e)$ satisfying the conditions:

Injectivity: If $f, g \in S_1$ and $C_f = C_g$, then $f = g$.

Permanence of the most rudimentary functional equations: If f and af belong to S_1 (where a is a complex number), then $C_{af} = aC_f$. If $f, g \in S_1$ and $f + g \in S_1$, then $C_{f+g} = C_f + C_g$.

Compatibility : The continuation $f \to C_f$ compatible with analytic continuation in the sense that if f has an analytic continuation across a boundary point $e^{i\theta}$, then it must agree with C_f near $e^{i\theta}$.

DEFINITION 9.1.1. A mapping $f \to C_f$ satisfying the above three properties is called a 'generalized analytic continuation' (GAC).

REMARK 9.1.2.   (1) The notion of GAC was formally studied by one of the present authors in [**128**].
   (2) The above three conditions seem to be the essential *minimum* needed to obtain a meaningful notion of 'continuation'. There are several others one might consider: *Preservation of products*: If $f, g \in S_1$ and $fg \in S_1$, then $C_{fg} = C_f C_g$. *Differentiability*: If $f, f' \in S_1$, then $C_{f'} = C'_f$. *Symmetry*: The mapping $C_f \to f$ is a GAC.

Using our previous work, especially the 'Abelian theorems' (Theorem 4.1.14, Theorem 5.2.3, Remark 6.2.2, and Theorem 7.3.1), one can show without much difficulty that the continuations mediated by Borel-Walsh, Gončar, Walsh-Tumarkin (pseudocontinuation), and almost periodic sequences, are in fact generalized analytic continuations.

Although the notion of GAC is very general, there are individual functions which cannot belong to any 'continuable' class $S_1 \subset \mathfrak{M}(\mathbb{D})$. For example, if f has an isolated winding point at $\zeta_0 \in \mathbb{T}$, then f cannot belong to any such class S_1, since the compatibility condition with analytic continuation is not satisfied (see Example 6.2.3).

Our new type of continuation, (FM) continuation (Chapter 8), has been put to good use in examining the cyclic vectors for the backward shift on various spaces of analytic functions. Nevertheless, it fails to be a GAC as in the above sense since, as we have seen (Theorem 8.5.1), (FM) continuations are not unique. Without uniqueness, the map which associates f with its (FM) continuation C_f is not even well defined. Nevertheless, as we shall see in a moment, (FM) continuation is indeed a generalized analytic continuation if one is willing to impose some restrictions.

9.2. Compatibility

In the theory of divergent series, there are important theorems which relate the various summability methods of Abel, Borel, Cesàro, and others. What is lacking here, and which we leave for future investigations, are the analogous theorems for GAC. We mention a few open questions in this regard.

QUESTION 9.2.1 (Gončar continuation vs. pseudocontinuation). Suppose $f \in \mathfrak{M}(\mathbb{D})$ has *both* a Gončar continuation $F_1 \in \mathfrak{M}(\mathbb{D}_e)$ *and* a pseudocontinuation $F_2 \in \mathfrak{M}(\mathbb{D}_e)$. Must $F_1 = F_2$? To better appreciate the question, consider the sum of any Borel series

$$\sum_{n=1}^{\infty} \frac{c_n}{z - z_n},$$

where $\{z_n\} \subset \mathbb{D}_e$ are distinct, all the cluster points of $\{z_n\}$ are on the unit circle, and the c_n's are non-zero and satisfy

$$\varlimsup_{n \to \infty} \sqrt[n]{|c_n|} < 1.$$

Considered initially for z in $\mathbb{D}$, this series defines a function $f \in \mathfrak{H}(\mathbb{D})$ which has a Gončar continuation in $\mathfrak{M}(\mathbb{D}_e)$. This Gončar continuation is equal to the sum of the above series on $\mathbb{D}_e$ and hence has a pole at each point z_n. Thus, if the sequence $\{z_n\}$ is chosen to accumulate non-tangentially at almost every point of $\mathbb{T}$, the Gončar continuation will not have finite non-tangential limit values almost everywhere, and so cannot be the pseudocontinuation of any function in $\mathfrak{M}(\mathbb{D})$ (see Proposition 6.6.4 and the discussion that follows). However, this does not imply a negative answer to Question 9.2.1 because f may possess no pseudocontinuation at all. Thus, it is essential in Question 9.2.1 that we postulate the existence of both a Gončar continuation *and* a pseudocontinuation. So far as we know, it is an open question as to whether or not these two continuations must be equal.

QUESTION 9.2.2 (Bochner-Bohnenblust vs. pseudocontinuation). The function

$$f = \sum_{n=1}^{\infty} \frac{c_n}{1 - e^{i\theta_n}z}, \quad z \in \mathbb{D},$$

(where $\{c_n\}$ is an absolutely summable sequence of non-zero complex numbers and $\{e^{i\theta_n}\}$ is a dense set of distinct points in $\mathbb{T}$) given by Poincaré, does not have an analytic continuation across any point of the circle (Chapter 3). However, it does have a pseudocontinuation across $\mathbb{T}$. As we have seen in Chapter 7, this Poincaré example is a special case of a more general continuation $f_A \to F_A$ involving almost periodic sequences. Is this continuation really new? That is to say, is there a function f_A which does not have a pseudocontinuation across any arc of $\mathbb{T}$? Supposing that the continuation $f_A \to F_A$ is not mediated by pseudocontinuation (which presumably is possible, although we do not know of such an example), is this type of continuation at least compatible with pseudocontinuation? More precisely, if f_A has a pseudocontinuation, must it be equal to F_A?

QUESTION 9.2.3 (Formal multiplication of series (FM)). Our new type of continuation, by formal multiplication of series (FM) described in Chapter 8, is not, in general, a GAC because of cases where uniqueness fails (Theorem 8.5.1). Furthermore, as seen in Example 8.5.5, a function $f \in \mathfrak{H}(\mathbb{D})$ might have two (FM) continuations, where one is a pseudocontinuation of f while the other is not. Thus, without uniqueness, questions about the compatibility of (FM) continuation with the other types of continuations, namely Gončar and pseudocontinuation, are pointless. However, we can combine our (FM) continuation with our discussion of the spectral properties of $T = B|\mathcal{M}$ (as in Chapter 8, § 8.4) to define a 'continuation' which is indeed a GAC. From here one can pose more meaningful compatibility questions.

Indeed, let $\mathcal{X}$ be a Banach space of analytic functions on $\mathbb{D}$ which satisfies our usual properties from Remark 8.4.23 (*e.g.*, the Dirichlet-type spaces D_α, the spaces ℓ_A^p). With these conditions, we were able to identify the parts of the spectrum of the operator $T = B|\mathcal{M}$, where $\mathcal{M}$ is a non-trivial B-invariant subspace of $\mathcal{X}$. In particular (Theorem 8.4.22), $\sigma(T) \cap \mathbb{D}$ is discrete if and only if $\text{ind}(\mathcal{M}^\perp) = 1$ and in this case

$$\sigma(T) \cap \mathbb{D} = \left\{ w : \frac{1}{1 - wz} \in \mathcal{M} \right\} = \{ \bar{\lambda} : \lambda \in Z(\mathcal{M}^\perp) \},$$

where $Z(\mathcal{M}^\perp)$ is the set of common zeros - in $\mathbb{D}$ - of the S-invariant subspace $\mathcal{M}^\perp \subset D_{-\alpha}$ (see eq.(8.4.18)). Moreover, in the case where $\sigma(T) \cap \mathbb{D}$ is discrete, the function

$$\lambda \to c_\lambda(f) = ((I - \lambda T)^{-1} f)(0),$$

is analytic on the set $\{1/\lambda : \lambda \notin \sigma(T)\}$ and equal to f whenever $\lambda \in \mathbb{D}$. In this sense, the function $\lambda \to c_\lambda(f)$, for $|\lambda| > 1$, can be regarded as a 'continuation' of f. In fact, as we have seen earlier, if $|\lambda| > 1$ with

$$1/\lambda \notin \sigma(T) \cap \mathbb{D} = \{ \bar{\lambda} : \lambda \in Z(\mathcal{M}^\perp) \}$$

then there is a $g \in \mathcal{M}^\perp$ such that $g(1/\bar{\lambda}) \neq 0$ and moreover,

$$(9.2.4) \qquad\qquad c_\lambda(f) = \left\langle \frac{zf}{z - \lambda}, g \right\rangle \Big/ -\bar{g}(1/\bar{\lambda}).$$

Furthermore, the quantity on the right hand side of the above equation is independent of such $g \in \mathcal{M}^\perp$ with $-g(1/\bar{\lambda}) \neq 0$ and is a FM continuation of f (Proposition 8.4.13 and Proposition 8.4.12), i.e.,

$$-\bar{g}(1/\bar{\lambda}) \# f = \left\langle \frac{zf}{z - \lambda}, g \right\rangle.$$

The mapping (continuation) from $\mathcal{M}$ to $\mathfrak{M}(\mathbb{D}_e)$ defined by

$$f \to c_\lambda(f)$$

is well defined (assuming $\sigma(T) \cap \mathbb{D}$ is discrete) and linear. Furthermore (Theorem 8.2.1), this continuation is compatible with analytic continuation in the sense that if f has an analytic continuation across $e^{i\theta}$ to an open neighborhood U containing $e^{i\theta}$, then $c_\lambda(f) = f(\lambda)$ for all $\lambda \in U \cap \mathbb{D}_e$. To show that $f \to c_\lambda(f)$ is a generalized analytic continuation, as defined in Definition 9.1.1, we need to check this map is injective. The theorem here is the following:

THEOREM 9.2.5. *Let $\mathcal{X}$ be as above and $\mathcal{M}$ be a non-trivial B-invariant subspace of $\mathcal{X}$ such that*

$$\mathcal{M}^\perp = \bigvee \{ z^n g : n = 0, 1, 2, \ldots \}^{\cdot}$$

for some $g \in \mathcal{X}^*$. Then $\sigma(T) \cap \mathbb{D}$ is discrete and the mapping from $\mathcal{M}$ to $\mathfrak{M}(\mathbb{D}_e)$ defined by $f \to c_\lambda(f)$ is injective and hence a generalized analytic continuation.

PROOF. The discreteness of $\sigma(T) \cap \mathbb{D}$ follows from Example 8.4.16 and Theorem 8.4.22. As mentioned in the discussion above, we just need to check that the mapping $f \to c_\lambda(f)$ is injective. Indeed, suppose that for a fixed $f \in \mathcal{M}$ the function $\lambda \to c_\lambda(f)$ is identically equal to zero on $\mathbb{D}_e$. It follows from eq.(9.2.4) that

$$\left\langle \frac{f}{z - \lambda}, h \right\rangle = 0$$

for every $h \in \mathcal{M}^\perp$ and $|\lambda| > 1$.

A computation with power series shows that if $f = \sum_k a_k z^k$, then

$$\frac{f}{z-\lambda} = -\sum_{l=0}^{\infty} z^l \left(\sum_{k=0}^{l} \frac{a_{l-k}}{\lambda^{k+1}} \right).$$

Thus if $h = \sum_l b_l z^l$, then

$$\left\langle \frac{f}{z-\lambda}, h \right\rangle = -\sum_{l=0}^{\infty} \left(\sum_{k=0}^{l} \frac{a_{l-k}}{\lambda^{k+1}} \right) \overline{b_l},$$

which, after reversing the order of summation, yields

$$\left\langle \frac{f}{z-\lambda}, h \right\rangle = -\sum_{k=0}^{\infty} \frac{1}{\lambda^{k+1}} \left(\sum_{l=k}^{\infty} a_{l-k} \overline{b_l} \right) = -\sum_{k=0}^{\infty} \frac{1}{\lambda^{k+1}} \langle z^k f, h \rangle.$$

Since we are assuming that the left-hand side of the above is zero for all $|\lambda| > 1$ and for all $h \in \mathcal{M}^{\perp}$, we conclude that $z^k f \in \mathcal{M}$ for each $k = 0, 1, 2, \ldots$

An argument used in proving Proposition 8.3.4 shows that if g is our generating vector for $\mathcal{M}^{\perp}$ (as in the hypothesis of the theorem) and $G(z) := \overline{g}(1/\overline{z})$ then, for $q \in \mathcal{X}$,

$$q \in \mathcal{M} \Leftrightarrow q \# G \text{ is analytic on } \mathbb{D}_e \text{ and vanishes at infinity.}$$

Also note that every $q \in \mathcal{M}$ has a FM continuation given by

$$\frac{q \# G}{G}$$

[1] and, due to the discreteness of $\sigma(T) \cap \mathbb{D}$ and hence the independence of $C_{q,h} = (q \# H)/H$ on $h \in \mathcal{M}^{\perp}$, $H(z) = \overline{h}(1/\overline{z})$, every $q \in \mathcal{M}$ has a unique continuation of the above type and moreover the pole at infinity of this continuation is at most a fixed number, say N, regardless of the $q \in \mathcal{M}$.

To show that f must be the zero function, thus showing the mapping

$$f \rightarrow c_{\lambda}(f)$$

is injective, we note from above that $z^k f \in \mathcal{M}$ for every $k = 0, 1, 2, \ldots$ For each k, $z^k f$ has a FM continuation of the type discussed in the previous paragraph given by

$$\frac{f \# G}{G/z^k},$$

which, for large enough k, will have a pole at infinity exceeding N, unless f is the zero function, which must therefore be the case. $\qquad \square$

REMARK 9.2.6.   (1) For the weighted Bergman spaces D_{α} ($\alpha < 0$), we can use results in [11] to show that for any non-trivial B-invariant subspace $\mathcal{M} \subset D_{\alpha}$, the set $\sigma(T) \cap \mathbb{D}$ is discrete and so the mapping $f \rightarrow c_{\lambda}(f)$ is well defined. We don't even need to use Theorem 9.2.5 to prove this mapping is injective (and hence a GAC). Indeed, the function $\lambda \rightarrow c_{\lambda}(f)$ is a pseudocontinuation of f [11, Thm. 4.8], and so the injectivity follows from the Lusin-Privalov uniqueness theorem.

---

[1]Note that the continuation $(q \# G)/G$ is just $C_{q,g}$.

(2) For the spaces D_α ($\alpha > 0$) however, the situation is more complicated. For a general B-invariant subspace $\mathcal{M}$ of D_α, the S-invariant subspace $\mathcal{M}^\perp \subset D_{-\alpha}$ might not have the division property (*i.e.*, $\mathcal{M}^\perp$ might satisfy the condition $\mathrm{ind}(\mathcal{M}^\perp) > 1$) making $\sigma(T) = \mathbb{D}^-$ and $\mathcal{M}^\perp$ not singly generated, defying the hypothesis of Theorem 9.2.5. The hypothesis that $\mathcal{M}^\perp$ is singly generated gives us both the discreteness of $\sigma(T) \cap \mathbb{D}$, allowing us to define the continuation in the first place, and enough technical conditions to make to make it a GAC. There is an alternative approach: If $0 < \alpha \leq 1$ and we assume that $\mathcal{M}^\perp$ has the division property, then, by remarkable results in [13] (for $\alpha = 1$) and [134] (for $0 < \alpha \leq 1$), $\mathcal{M}^\perp$ is singly generated. In this case, the hypothesis of the above theorem is satisfied and so the mapping $f \to c_\lambda(f)$ is a GAC. At this writing, it is not known if the condition that $\mathcal{M}^\perp$ has the division property implies that $\mathcal{M}^\perp$ is singly generated for S-invariant subspaces $\mathcal{M}^\perp$ of $D_{-\alpha}$ when α is outside $[0, 1]$.

(3) For certain weighted spaces, $\mathcal{M}^\perp$ is always singly generated and so the above theorem applies.

(4) Since $T = B|\mathcal{M}$, the definition of the function

$$\lambda \to c_\lambda(f) = ((I - \lambda T)^{-1} f)(0),$$

seems to depend on the B-invariant subspace $\mathcal{M}$ containing f. In a way, this is indeed the case. However, if $\mathcal{M}_1 \subset \mathcal{M}_2$ are both B-invariant subspaces of $\mathcal{X}$ such that both $\sigma(B|\mathcal{M}_1) \cap \mathbb{D}$ and $\sigma(B|\mathcal{M}_2) \cap \mathbb{D}$ are discrete, then it follows from Theorem 8.4.22 and Proposition 8.4.9 that

$$\sigma(B|\mathcal{M}_1) \subset \sigma(B|\mathcal{M}_2).$$

Letting $T_j = B|\mathcal{M}_j, j = 1, 2$, and

$$c_\lambda(f, T_j) = ((I - \lambda T_j)^{-1} f)(0) \text{ for } f \in \mathcal{M}_1,$$

we can use Proposition 8.4.13 to prove that for $f \in \mathcal{M}_1$, the function $\lambda \to c_\lambda(f, T_2)$ has an analytic continuation, given by $\lambda \to c_\lambda(f, T_1)$, to the set $\{1/\lambda : \lambda \notin \sigma(T_1)\}$.

QUESTION 9.2.7 (Tauberian theorems). Pursuing the analogy with summability theory, the original and 'purest' Tauberian theorem[2], due to Tauber, examines conditions on the terms of a series in the presence of which Abel summability implies convergence.

This suggests the study of questions of the following kind: Suppose f is holomorphic (meromorphic) in a domain G and has a 'continuation' C_f (by some specified type of GAC) that is holomorphic (meromorphic) in some domain G'. We will suppose G and G' are disjoint but have a common smooth boundary arc J. Are there ('Tauberian') conditions we can place on f, or perhaps on the pair f, C_f, which then ensure that f and C_f are actually analytic continuations of one another?

If the type of GAC considered is pseudocontinuation, one might suppose some sufficient amount of smoothness of f on J would suffice for this. However, we do not know of any such result short of the trivial hypothesis that f allows analytic continuation, and then this is just the compatibility result. The condition that f is

---

[2]If $\sum a_n$ is Abel summable to s and $a_n = o(1/n)$, then $\sum a_n = s$ (see [68, p. 149]).

holomorphic with C^∞-regularity up to the boundary is not enough, as can be seen with the following counterexample when G is the unit disk: If ϕ denotes the atomic inner function

$$\phi(z) = \exp\left(\frac{z+1}{z-1}\right),$$

then $(\phi H^2)^\perp$ contains non-trivial functions f whose derivatives (of all orders) extend continuously to $\mathbb{D}^-$. Indeed, we can take f to be $P(\phi(e^{it})e^{-it}\overline{u}(e^{it}))$, where P is the Riesz projection of $L^2(\mathbb{T})$ onto $H^2(\mathbb{T})$ (see eq.(6.8.15)), and u is any (non-trivial) holomorphic function in $\mathbb{D}$ whose derivatives, of all orders, extend continuously to $\mathbb{D}^-$, and vanishing to infinite order at $z = 1$. Such a function f has an analytic continuation to $\mathbb{C}_\infty\backslash\{1\}$ [**51**, Cor. 3.1.10] but will not have an analytic continuation across $z = 1$. So, if there is a nontrivial result to be found here, the appropriate regularity to impose on the boundary function for f will be something between C^∞ and analyticity, like a Gevrey condition (see [**66**, p. 27] or [**138**, p. 25] for a definition). Investigations providing some information in this general direction, but which do not answer the question just raised, can be found in [**125, 130**]. If we are willing to impose conditions on *both* f and C_f there is a classical result [**60**, p. 95] (sort of a 'Morera's theorem'): if f and C_f are holomorphic and of Hardy class H^1 for their respective domains and C_f is a pseudocontinuation of f across J, then C_f is an analytic continuation of f across J. Moreover, the H^1 hypothesis is essentially the weakest that will allow such a conclusion. The same questions, but with Gončar continuation or FM continuation replacing pseudocontinuation in the hypothesis, seem to be totally uncharted territory.

List of Symbols

139

Bibliography

1. E. Abakumov, *Essais sur les operateurs de Hankel et capacité d'approximation des séries lacunaires*, Ph.D., L'Universite Bordeaux I, May 1994.
2. _____, *Cyclicity and approximation by lacunary power series*, Michigan Math. J. **42** (1995), no. 2, 277–299. MR **96f**:47056
3. N. I. Akhiezer, *The classical moment problem and some related questions in analysis*, Hafner Publishing Co., New York, 1965. MR 32 #1518
4. A. B. Aleksandrov, *Invariant subspaces of the backward shift operator in the space H^p ($p \in (0, 1)$)*, Zap. Nauchn. Sem. Leningrad. Otdel. Mat. Inst. Steklov. (LOMI) **92** (1979), 7–29, 318, Investigations on linear operators and the theory of functions, IX. MR **81h**:46018
5. _____, *Invariant subspaces of shift operators. An axiomatic approach*, Zap. Nauchn. Sem. Leningrad. Otdel. Mat. Inst. Steklov. (LOMI) **113** (1981), 7–26, 264, Investigations on linear operators and the theory of functions, XI. MR **83g**:47031
6. _____, *Invariant subspaces of the backward shift operator in the Smirnov class*, Lecture Notes in Math., vol. 1043, pp. 393 – 395, Springer-Verlag, Berlin-New York, 1984.
7. _____, *Lacunary series and pseudocontinuations*, Zap. Nauchn. Sem. S.-Peterburg. Otdel. Mat. Inst. Steklov. (POMI) **232** (1996), no. Issled. po Linein. Oper. i Teor. Funktsii. 24, 16–32, 213. MR **98j**:30001
8. _____, *Gap series and pseudocontinuations. An arithmetic approach*, Algebra i Analiz **9** (1997), no. 1, 3–31. MR **98d**:30006
9. A. Aleman, *Hilbert spaces of analytic functions between the Hardy and the Dirichlet space*, Proc. Amer. Math. Soc. **115** (1992), no. 1, 97–104. MR **92i**:46030
10. A. Aleman and S. Richter, *Simply invariant subspaces of H^2 of some multiply connected regions*, Integral Equations Operator Theory **24** (1996), no. 2, 127–155. MR **99b**:47010a
11. A. Aleman, S. Richter, and W. T. Ross, *Pseudocontinuations and the backward shift*, Indiana Univ. Math. J. **47** (1998), no. 1, 223–276. MR **2000i**:47009
12. A. Aleman, S. Richter, and C. Sundberg, *The majorization function and the index of invariant subspaces in the Bergman spaces*, to appear: J. Anal. Math.
13. _____, *Beurling's theorem for the Bergman space*, Acta Math. **177** (1996), no. 2, 275–310. MR **98a**:46034
14. A. Aleman and W. T. Ross, *The backward shift on weighted Bergman spaces*, Michigan Math. J. **43** (1996), no. 2, 291–319. MR **97i**:47053
15. D. Z. Arov, *Darlington's method in the study of dissipative systems*, Dokl. Akad. Nauk SSSR **201** (1971), no. 3, 559–562. MR 55 #1127
16. F. Bagemihl and W. Seidel, *Some boundary properties of analytic functions*, Math. Z. **61** (1954), 186–199. MR 16,460d
17. H. Bercovici, C. Foiaş, and C. Pearcy, *Dual algebras with applications to invariant subspaces and dilation theory*, Published for the Conference Board of the Mathematical Sciences, Washington, DC, 1985. MR **87g**:47091
18. A. Beurling, *Sur les fonctions limites quasi analytiques des fractions rationnelles*, Proc. Eighth Scand. Math. Congress, Lund (1935), 199–210.
19. _____, *Ensembles exceptionnels*, Acta Math. **72** (1939), 1–13. MR 1,226a
20. _____, *Un théorème sur les fonctions bornées et uniformément continues sur l'axe réel*, Acta Math. **77** (1945), 127–136. MR 7,61f
21. _____, *On two problems concerning linear transformations in Hilbert space*, Acta Math. **81** (1948), 17. MR 10,381e

22. _____, *The collected works of Arne Beurling. Vol. 1*, Birkhäuser Boston Inc., Boston, MA, 1989, Complex analysis, Edited by L. Carleson, P. Malliavin, J. Neuberger and J. Wermer. MR **92k**:01046a

23. R. Boas, *Entire functions*, Academic Press Inc., New York, 1954. MR 16,914f

24. S. Bochner and F. Bohnenblust, *Analytic functions with almost periodic coefficients*, Ann. Math. **35** (1934), 152–161.

25. H. Bohr, *Almost Periodic Functions*, Chelsea Publishing Company, New York, N.Y., 1947. MR 8,512a

26. F. F. Bonsall, *Decompositions of functions as sums of elementary functions*, Quart. J. Math. Oxford Ser. (2) **37** (1986), no. 146, 129–136. MR **87h**:46090

27. É. Borel, *Leçons sur les fonctions monogènes uniformes d'une variable complexe*, Gauthier-Villars, Paris, 1917.

28. _____, *Leçons sur la Théorie des Fonctions*, third ed., Gauthier-Villars, Paris, 1928.

29. P. Bourdon, *Cellular-indecomposable operators and Beurling's theorem*, Michigan Math. J. **33** (1986), no. 2, 187–193. MR **87g**:47048

30. P. Bourdon and J. Shapiro, *Spectral synthesis and common cyclic vectors*, Michigan Math. J. **37** (1990), no. 1, 71–90. MR **91m**:47039

31. L. Brown and A. L. Shields, *Cyclic vectors in the Dirichlet space*, Trans. Amer. Math. Soc. **285** (1984), no. 1, 269–303. MR **86d**:30079

32. L. Brown, A. L. Shields, and K. Zeller, *On absolutely convergent exponential sums*, Trans. Amer. Math. Soc. **96** (1960), 162–183. MR 26 #332

33. R. Carmichael, *Linear differential equations of infinite order*, Bull. Amer. Math. Soc. **42** (1936), 193–218.

34. _____, *On non-homogeneous linear differential equations of infinite order with constant coefficients*, Amer. J. Math. **58** (1936), 473–486.

35. J. A. Cima and W. T. Ross, *The backward shift on the Hardy space*, American Mathematical Society, Providence, RI, 2000.

36. E. F. Collingwood and A. J. Lohwater, *The theory of cluster sets*, Cambridge University Press, Cambridge, 1966. MR 38 #325

37. J. B. Conway, *A course in functional analysis*, Springer-Verlag, New York, 1985. MR **86h**:46001

38. C. Corduneanu, *Almost periodic functions*, Interscience Publishers, New York-London-Sydney, 1968. MR 58 #2006

39. H. Cramér, *Un théorème sur les séries de Dirichlet et son application*, Ark. Mat. Astr. Fys. **13** (1918), 1–14.

40. L. de Branges and J. Rovnyak, *Canonical models in quantum scattering theory*, Perturbation Theory and its Applications in Quantum Mechanics (Proc. Adv. Sem. Math. Res. Center, U.S. Army, Theoret. Chem. Inst., Univ. of Wisconsin, Madison, Wis., 1965), Wiley, New York, 1966, pp. 295–392. MR 39 #6109

41. J. Delsarte, *Les fonctions moyenne-périodiques*, J. de Math. Pures et Appliqués **14** (1935), 403 – 453.

42. A. Denjoy, *Sur les séries de fractions rationnelles*, Bull. Soc. Math. France **52** (1924), 418–434.

43. _____, *Sur les singularités des séries de fractions rationnelles*, Rend. Circ. Mat. Palermo **50** (1926), 1–95.

44. P. Dewilde, *Generalized Darlington synthesis*, IEEE Trans. Circuits Systems I Fund. Theory Appl. **46** (1999), no. 1, 41–58. MR **2000d**:93027

45. P. Dienes, *The Taylor series: An introduction to the theory of functions of a complex variable*, Dover Publications Inc., New York, 1957. MR 19,735d

46. A. E. Djrbashian and F. A. Shamoian, *Topics in the theory of A_α^p spaces*, BSB B. G. Teubner Verlagsgesellschaft, Leipzig, 1988. MR **91k**:46019

47. Y. Domar, *On spectral analysis in the narrow topology*, Math. Scand. **4** (1956), 328–332. MR 19,413b

48. _____, *On the existence of a largest subharmonic minorant of a given function*, Ark. Mat. **3** (1957), 429–440. MR 19,408c

49. R. G. Douglas and J. W. Helton, *Inner dilations of analytic matrix functions and Darlington synthesis*, Acta Sci. Math. (Szeged) **34** (1973), 61–67. MR 48 #900

50. R. G. Douglas, H. S. Shapiro, and A. L. Shields, *On cyclic vectors of the backward shift*, Bull. Amer. Math. Soc. **73** (1967), 156–159. MR 34 #3316

51. ———, *Cyclic vectors and invariant subspaces for the backward shift operator.*, Ann. Inst. Fourier (Grenoble) **20** (1970), no. fasc. 1, 37–76. MR 42 #5088

52. P. L. Duren, *Theory of H^p spaces*, Academic Press, New York, 1970. MR 42 #3552

53. P. L. Duren, B. W. Romberg, and A. L. Shields, *Linear functionals on H^p spaces with $0 < p < 1$*, J. Reine Angew. Math. **238** (1969), 32–60. MR 41 #4217

54. P. Erdős, *Note on the converse of Fabry's gap theorem*, Trans. Amer. Math. Soc. **57** (1945), 102–104. MR 6,148f

55. E. Fabry, *Sur les séries de Taylor qui ont une infinité de points singuliers*, Acta Math. **22** (1898–1899), 65–88.

56. P. Fatou, *Séries trigonométriques et séries de Taylor*, Acta Math. **30** (1906), 335 – 400.

57. C. Foiaş, *A remark on the universal model for contractions of G. C. Rota*, Com. Acad. R. P. Romîne **13** (1963), 349–352. MR 31 #605

58. I. Fredholm, *Om en speciell klass av singulära linjer*, Öfv. av K. Svenska Vet-Akad Förh **47** (1890), 131–134.

59. ———, *Œuvres complètes de Ivar Fredholm*, Kungl. Svenska Vetensapsakademien, Djursholm, Sweden, 1955.

60. J. B. Garnett, *Bounded analytic functions*, Academic Press Inc., New York, 1981. MR **83g**:30037

61. A. O. Gel'fond, *Ischislenie konechnykh raznostei*, Gosudarstv. Izdat. Fiz.-Mat. Lit., Moscow, 1959. MR 35 #7020

62. A. A. Gončar, *On quasianalytic continuation of analytic functions across a Jordan arc*, Dokl. Akad. Nauk SSSR **166** (1966), 1028–1031. MR 34 #320

63. ———, *Generalized analytic continuation*, Mat. Sb. (N.S.) **76 (118)** (1968), 135–146. MR 38 #323

64. M. B. Gribov and N. K. Nikol'skiĭ, *Invariant subspaces and rational approximation*, Zap. Nauchn. Sem. Leningrad. Otdel. Mat. Inst. Steklov. (LOMI) **92** (1979), 103–114, 320, Investigations on linear operators and the theory of functions, IX. MR **81f**:47006

65. J. Hadamard, *Essai sur l'étude des fonctions données par leur développement de Taylor*, J. Math. **8** (1892), 101–186.

66. ———, *Lectures on Cauchy's problem in linear partial differential equations*, Dover Publications, New York, 1953. MR 14,474f

67. G. H. Hardy, *Weierstrass's non-differentiable function*, Trans. Amer. Math. Soc. **17** (1916), no. 3, 301–325.

68. ———, *Divergent Series*, Oxford, at the Clarendon Press, 1949. MR 11,25a

69. V. Havin and B. Jöricke, *The uncertainty principle in harmonic analysis*, Springer-Verlag, Berlin, 1994. MR **96c**:42001

70. H. Hedenmalm, *An invariant subspace of the Bergman space having the codimension two property*, J. Reine Angew. Math. **443** (1993), 1–9. MR **94k**:30092

71. H. Hedenmalm, B. Korenblum, and K. Zhu, *Theory of Bergman spaces*, Springer-Verlag, New York, 2000. MR **2001c**:46043

72. H. Hedenmalm, S. Richter, and K. Seip, *Interpolating sequences and invariant subspaces of given index in the Bergman spaces*, J. Reine Angew. Math. **477** (1996), 13–30. MR **97i**:46044

73. H. M. Hilden and L. J. Wallen, *Some cyclic and non-cyclic vectors of certain operators*, Indiana Univ. Math. J. **23** (1973/74), 557–565. MR 48 #4796

74. E. Hille, *Dirichlet series with complex exponents*, Ann. Math. **25** (1924), no. 3, 261–278.

75. ———, *Analytic function theory. Vol. II*, Ginn and Co., Boston, Mass.-New York-Toronto, Ont., 1962. MR 34 #1490

76. D. Hitt, *Invariant subspaces of H^2 of an annulus*, Pacific J. Math. **134** (1988), no. 1, 101–120. MR **90a**:46059

77. K. Hoffman, *Banach spaces of analytic functions*, Dover Publications Inc., New York, 1988, Reprint of the 1962 original. MR **92d**:46066

78. K. Hoffman and R. Kunze, *Linear algebra*, Prentice-Hall Inc., Englewood Cliffs, N.J., 1971. MR 43 #1998

79. C. Horowitz, *Zeros of functions in the Bergman spaces*, Duke Math. J. **41** (1974), 693–710. MR 50 #10215

80. B. Jöricke, *On pseudoanalytic continuation and the behavior of the boundary values of analytic functions on small sets*, preprint P-MATH-22/81, Inst. für Mathematik, Akad. der Wiss. der DDR, Berlin, 1981.

81. ———, *Pseudocontinuability and properties of analytic functions on boundary sets of positive measure*, J. Soviet Math. **27** (1984), no. 1, 2481–2486.

82. J.-P. Kahane, *Sur quelques problèmes d'unicité et de prolongement, relatifs aux fonctions approchables par des sommes d'exponentielles*, Ann. Inst. Fourier, Grenoble **5** (1953–1954), 39–130 (1955). MR 17,732b

83. ———, *Lecture notes on mean periodic functions*, Tata Institute, Bombay, 1959.

84. ———, *Lacunary Taylor and Fourier series*, Bull. Amer. Math. Soc. **70** (1964), 199–213. MR 29 #223

85. J.-P. Kahane and Y. Katznelson, *Sur le comportement radial des fonctions analytiques*, C. R. Acad. Sci. Paris Sér. A-B **272** (1971), A718–A719. MR 43 #3454

86. V. E. Katsnelson, *Description of a class of functions which admit an approximation by rational functions with preassigned poles. I*, Matrix and operator valued functions, Birkhäuser, Basel, 1994, pp. 87–132. MR **96e**:30095

87. Y. Katznelson, *An introduction to harmonic analysis*, corrected ed., Dover Publications Inc., New York, 1976. MR 54 #10976

88. D. Khavinson and H. S. Shapiro, *The heat equation and analytic continuation: Ivar Fredholm's first paper*, Exposition. Math. **12** (1994), no. 1, 79–95. MR **95b**:35002

89. D. Khavinson and M. Stessin, *Certain linear extremal problems in Bergman spaces of analytic functions*, Indiana Univ. Math. J. **46** (1997), no. 3, 933–974. MR **99k**:30080

90. P. Koosis, *The logarithmic integral. I*, Cambridge University Press, Cambridge, 1988. MR **90a**:30097

91. ———, *Introduction to H_p spaces*, second ed., Cambridge University Press, Cambridge, 1998. MR **2000b**:30052

92. T. L. Kriete, III, *On the Fourier coefficients of outer functions*, Indiana Univ. Math. J. **20** (1970/71), 147–155. MR 54 #555

93. A. F. Leont'ev, *Differential equations of infinite order and their applications*, Proc. Fourth All-Union Math. Congr. (Leningrad, 1961) (Russian), Vol. II, Izdat. "Nauka", Leningrad, 1964, pp. 648–660. MR 36 #2920

94. T. A. Leont'eva, *The representation of analytic functions by series of rational functions*, Mat. Zametki **2** (1967), 347–356. MR 36 #3995

95. B. Ya. Levin, *Lectures on entire functions*, American Mathematical Society, Providence, RI, 1996. MR **97j**:30001

96. B. M. Levitan and V. V. Zhikov, *Almost periodic functions and differential equations*, Cambridge University Press, Cambridge, 1982. MR **84g**:34004

97. W. Maak, *Fastperiodische Funktionen*, Springer-Verlag, Berlin, 1950. MR 13,29f

98. B. Malgrange, *Existence et approximation des solutions des équations aux dérivées partielles et des équations de convolution*, Ann. Inst. Fourier, Grenoble **6** (1955–1956), 271–355. MR 19,280a

99. A. I. Markushevich, *Theory of functions of a complex variable. Vol. I, II, III*, English ed., Chelsea Publishing Co., New York, 1977. MR 56 #3258

100. J. W. Moeller, *On the spectra of some translation invariant spaces*, J. Math. Anal. Appl. **4** (1962), 276–296. MR 27 #588

101. H. Muggli, *Differentialgleichungen unendlich hoher Ordnung mit konstanten Koeffizienten*, Comment. Math. Helv. **11** (1938), 151–179.

102. R. Newcombe, *Linear multiport synthesis*, McGraw Hill, New York, 1966.

103. N. K. Nikol'skiĭ, *Treatise on the shift operator*, Springer-Verlag, Berlin, 1986. MR **87i**:47042

104. I. Niven, H. S. Zuckerman, and H. L. Montgomery, *An introduction to the theory of numbers*, fifth ed., John Wiley & Sons Inc., New York, 1991. MR **91i**:11001

105. A. A. Pankov, *Bounded and almost periodic solutions of nonlinear operator differential equations*, Kluwer Academic Publishers Group, Dordrecht, 1990. MR **92f**:35002

106. H. Poincaré, *Sur les fonctions à espaces lacunaires*, Acta Soc. Scient. Fennicae **12** (1883), 341–350.

107. G. Pólya, *Eine Verallgemeinerung des Fabryschen Lückensatzes*, Nachr. Ges. Wiss. Göttingen (1927), 187–195.

108. ———, *On converse gap theorems*, Trans. Amer. Math. Soc. **52** (1942), 65–71. MR 4,7g

109. G. Pólya and G. Szegö, *Problems and theorems in analysis. I*, Springer-Verlag, Berlin, 1998, Reprint of the 1978 English translation.

110. I. I. Privalov, *Randeigenschaften analytischer Funktionen*, VEB Deutscher Verlag der Wissenschaften, Berlin, 1956. MR 18,727f

111. S. Richter, *Invariant subspaces in Banach spaces of analytic functions*, Trans. Amer. Math. Soc. **304** (1987), no. 2, 585–616. MR **88m**:47056

112. S. Richter and A. L. Shields, *Bounded analytic functions in the Dirichlet space*, Math. Z. **198** (1988), no. 2, 151–159. MR **89c**:46039

113. S. Richter and C. Sundberg, *Multipliers and invariant subspaces in the Dirichlet space*, J. Operator Theory **28** (1992), no. 1, 167–186. MR **95e**:47007

114. _____, *Invariant subspaces of the Dirichlet shift and pseudocontinuations*, Trans. Amer. Math. Soc. **341** (1994), no. 2, 863–879. MR **94d**:47026

115. F. Riesz, *Les Systèmes d'équations linéaires à une infinité d'inconnues*, Gauthier-Villars, Paris, 1913.

116. J. F. Ritt, *On a general class of linear homogeneous differential equations of infinite order with constant coefficients*, Trans. Amer. Math. Soc. **18** (1917), no. 1, 27–49. MR 1 501 060

117. _____, *Note on Dirichlet series with complex exponents*, Ann. Math. **26** (1924), no. 1/2, 144.

118. M. Rosenblum and J. Rovnyak, *Hardy classes and operator theory*, The Clarendon Press Oxford University Press, New York, 1985, Oxford Science Publications. MR **87e**:47001

119. G.-C. Rota, *On models for linear operators*, Comm. Pure Appl. Math. **13** (1960), 469–472. MR 22 #2898

120. H. L. Royden, *Invariant subspaces of H^p for multiply connected regions*, Pacific J. Math. **134** (1988), no. 1, 151–172. MR **90a**:46056

121. D. Sarason, *The H^p spaces of an annulus*, Mem. Amer. Math. Soc. No. **56** (1965), 78. MR 32 #6256

122. L. Schwartz, *Théorie générale des fonctions moyenne-périodiques*, Ann. of Math. (2) **48** (1947), 857–929. MR 9,428c

123. K. Seip, *Beurling type density theorems in the unit disk*, Invent. Math. **113** (1993), no. 1, 21–39. MR **94g**:30033

124. _____, *Regular sets of sampling and interpolation for weighted Bergman spaces*, Proc. Amer. Math. Soc. **117** (1993), no. 1, 213–220. MR **93c**:30051

125. F. A. Shamoyan, *On weak invertibility in weighted spaces of analytic functions*, Izv. Ross. Akad. Nauk Ser. Mat. **60** (1996), no. 5, 191–212. MR **97m**:46086

126. H. S. Shapiro, *The expansion of mean-periodic functions in series of exponentials.*, Comm. Pure Appl. Math. **11** (1958), 1–21. MR 21 #2157

127. _____, *Weakly invertible elements in certain function spaces, and generators in ℓ_1*, Michigan Math. J. **11** (1964), 161–165. MR 29 #3620

128. _____, *Generalized analytic continuation*, Symposia on Theoretical Physics and Mathematics, Vol. 8 (Symposium, Madras, 1967), Plenum, New York, 1968, pp. 151–163. MR 39 #2953

129. _____, *Overconvergence of sequences of rational functions with sparse poles.*, Ark. Mat. **7** (1968), 343–349. MR 38 #4658

130. _____, *Smoothness of the boundary function of a holomorphic function of bounded type*, Ark. Mat. **7** (1968), 443–447 (1968). MR 38 #4691

131. _____, *Functions nowhere continuable in a generalized sense*, Publ. Ramanujan Inst. No. **1** (1968/1969), 179–182. MR 42 #1982

132. H. S. Shapiro and A. L. Shields, *On the zeros of functions with finite Dirichlet integral and some related function spaces*, Math. Z. **80** (1962), 217–229. MR 26 #2617

133. S. Shimorin, *Approximate spectral synthesis in the Bergman space*, Duke Math. J. **101** (2000), no. 1, 1–39. MR **2001d**:47015

134. _____, *Wold-type decompositions and wandering subspaces for operators close to isometries*, J. Reine Angew. Math. **531** (2001), 147–189.

135. V. I. Smirnov, *Sur les valeurs limites des fonctions, régulières à l'intérieur d'un cercle*, Journal de la Société Phys.-Math. de Léningrade **2** (1929), 22–37.

136. C. Sundberg, *Analytic continuability of Bergman inner functions*, Michigan Math. J. **44** (1997), no. 2, 399–407. MR **98h**:46022

137. E. Titchmarsh, *The theory of functions*, Oxford University Press, Oxford, 1939.

138. F. Trèves, *Basic linear partial differential equations*, Academic Press, New York-London, 1975, Pure and Applied Mathematics, Vol. 62. MR 56 #6063

139. M. Tsuji, *Potential theory in modern function theory*, Chelsea Publishing Co., New York, 1975, Reprinting of the 1959 original. MR 54 #2990

140. G. C. Tumarkin, *Description of a class of functions admitting an approximation by fractions with preassigned poles*, Izv. Akad. Nauk Armjan. SSR Ser. Mat. **1** (1966), no. 2, 89–105. MR 34 #6123

141. G. Valiron, *Sur les solutions des équations différentielles linéaires d'ordre infini et à coéfficients constants*, Ann. de L'Ecole Normale Sup. **46** (1929), no. 3, 24–53.

142. H. von Koch, *Sur les systèmes d'une infinité d'équations linéaires à une infinité d'inconnues*, Proc. First Scand. Math. Congr. Stockholm, pp. 43–61, B. G. Teubner, Leipzig u. Berlin, 1910.

143. J. L. Walsh, *Interpolation and approximation by rational functions in the complex plane*, Amer. Math. Soc. Coll. Pub. (20), Providence, RI, 1935.

144. K. Weierstrass, *Über continuirliche Functionen eines reellen Arguments, die für keinen Werth des letzteren einen bestimmten Differentialquotienten besitzen*, Königl. Akad. Wiss., Mathematische Werke II (1872), 71–74.

145. W. Wogen, *On some operators with cyclic vectors*, Indiana Univ. Math. J. **27** (1978), no. 1, 163–171. MR 57 #7234

146. J. Wolff, *Sur les séries $\sum A_k/(z - z_k)$*, Comptes Rendus **173** (1921), 1057–1058, 1327–1328.

147. D. V. Yakubovich, *Invariant subspaces of the operator of multiplication by z in the space E^p in a multiply connected domain*, Zap. Nauchn. Sem. Leningrad. Otdel. Mat. Inst. Steklov. (LOMI) **178** (1989), no. Issled. Linein. Oper. Teorii Funktsii. 18, 166–183, 186–187. MR **91c**:47061

Index

MARSTON SCIENCE LIBRARY

Date Due

Due	Returned	Due	Returned